2022-2032 | INTERNATIONAL DECADE OF
Indigenous Languages

ANTICUS MULTICULTURAL ASSOCIATION

The 3rd Annual Kurultai of the Endangered Cultural Heritage
AKECH 2020
27-28 November 2020, Constanta, Romania, FULLY VIRTUAL
CONFERENCE PROCEEDINGS

2022-2032 | INTERNATIONAL DECADE OF
Indigenous Languages

Anticus Multicultural Association

The Annual Kurultai of the Endangered Cultural Heritage
AKECH

27-28 November 2020, Constanta, Romania, FULLY VIRTUAL
CONFERENCE PROCEEDINGS

Issue 3

ISSN 2668-3474, ISSN-L 2668-3474

Anticus Press, Constanța, 2020

Asociația Multiculturală „Anticus"
Editura Anticus Press
Constanța
www.anticusmulticultural.org
friends@anticusmulticultural.org

The 3rd Annual Kurultai of the Endangered Cultural Heritage
AKECH 2020, 27-28 November 2020, Constanta, Romania, Fully virtual

Co-Chairs:
Senior Lecturer Dr. Alan Reed Libert, Newcastle University, NSW, Australia
Dr. Maria Magdolna Tatár, retired from the University of Oslo

---

Journal: *The Annual Kurultai of the Endangered Cultural Heritage-AKECH*, Conference proceedings, ISSN 2668-3474, ISSN-L 2668-3474
Editor in Chief: Senior Lecturer Dr. Alan Reed Libert
Editors: Prof. dr. Kurmanbek Kyyanovich Abdyldaev, Dr. Maria Magdolna Tatár, Bakhtygul Makhanbetova, Taner Murat
Computerized editing: Elif Güney
Address: Luntrașului 16, 900388, Constanța, Romania
Phone: +40 961 273219
Web: www.anticusmulticultural.org
Email: friends@anticusmulticultural.org

Servind scriitorul, salvăm patrimoniul
Serving the writer, saving the heritage

Copyright © 2020
Toate drepturile asupra acestei ediții sunt rezervate autorului

Contents

Program 7

   The Annual Kurultai of the Endangered Cultural Heritage – AKECH, 3rd edition

Language Shift Reversal Thanks to COVID-19 9

   Valeria Buonocore

      Independent Scholar, Newcastle, Australia

Bilingual Language Behaviours in Multicultural Australia: An Insight into the Nature of Bilingualism and Directions for Sustainable Community Language Practices 18

   Dr. Iryna Khodos

      University of Newcastle, Newcastle, Australia

A Critical Examination of Teaching Methodology in ELICOS in Australia: A transition to online course under the COVID-19 crisis 31

   Sarah Mengshan Xu

      University of Newcastle, Australia

Distinct Linguistic Characteristics of Trakya Türkçesi: A Cultural Treasure .. 44

   Devrim Yılmaz

      University of New England, Armidale, Australia

Establishing Rapport with Evaluative Language in Online Hotel Responses 54

   Ly Wen Taw

      School of Humanities and Social Science, Faculty of Education and Arts, The University of Newcastle, Australia

      Department of English, Faculty of Modern Languages and Communication, Universiti Putra Malaysia, Malaysia

      Centre for the Advancement of Language Competence (CALC), Universiti Putra Malaysia, Malaysia

Distinguishing Between Language Difference and Language Disorder in Deaf Children who use Signed Language *66*

    Joanna Hoskin[1], Hilary Dumbrill[2], Wolfgang Mann[3]

    [1] City, University of London, UK

    [2] Hamilton Lodge School and College, Brighton, UK

    [3] University of Roehampton, UK

Index... *76*

Program

The Annual Kurultai of the Endangered Cultural Heritage – AKECH, 3rd edition

The Annual Lecture on Exile in Comparative Literature and the Arts – ALECLA, 3rd edition

Friday, November 27, 2020, 21:00 GMT (23:00 Bucharest)
Zoom Meeting

21:00 GMT
Opening and Greetings, Taner Murat

21:05 GMT
Grammatical Viruses and Real Ones: Implications for Endangered Languages, Senior Lecturer Dr. Alan Libert, University of Newcastle, Australia

Panel 1
Chair: Dr. Maria Magdolna Tatár, University of Oslo, retired

21:20 GMT
Blood and Honey: The Secret Herstory of Women: South Slavic Women's Experiences in a World of Modern-day Territorial Warfare, Dr. Danica Anderson

21:40 GMT
Power, Rule and Social Order in Traditional Moso Society, Ilaria Emma Borjigid Bohm, Xiamen University / University of Hamburg

22:00 GMT
Reclusive culture in Chinese Mountain and Water painting, Giacomo Bruni, China Academy of Art

22:20 GMT
Language shift reversal thanks to COVID-19, Valeria Buonocore

22:40 GMT
Bilingual language behaviours in multicultural Australia: An insight into the nature of bilingualism and directions for sustainable community language practices, Iryna Khodos, University of Newcastle, Australia

Saturday, November 28, 2020, 07:00 GMT (09:00 Bucharest)

Panel 2
Chair: Senior Lecturer Dr. Alan Libert

07:00 GMT
Indigenous Sign Languages, Gestures, and its Revitalisation: Implications for deaf and non-deaf Indigenous Communities in the Pacific, Rodney Adams, University of Newcastle

07:20 GMT
A Critical Examination of Teaching Methodology in ELICOS in Australia: a Transition to Online Course under the COVID-19 Crisis, Sarah Mengshan Xu, University of Newcastle, Australia

07:40 GMT
Distinct linguistic characteristics of Trakya Türkçesi: a cultural treasure, Devo Y. Devrim, University of New England, Australia

08:00 GMT
Cultural heritage and place naming: An ecolinguistic analysis of the interaction between humans and the environment, Dorcas Zuvalinyenga, University of Newcastle, Australia

08:20 GMT
Establishing Rapport with Evaluative Language in Online Hotel Responses, Ly Wen Taw, Universiti Putra Malaysia, Malaysia/The University of Newcastle, Australia

08:40 GMT
From the light of noon to the nocturnal earth. Gabriele d'Annunzio and the Franciscan nature
Donato Gagliastro, Institute of Greek and Latin Studies, Charles University, Prague, Czech Republic

09:00 GMT
Distinguishing between language difference and language disorder in deaf children who use sign language, Wolfgang Mann, School of Education, University of Roehampton & Hilary Dumbrill, Specialist Speech and Language Therapist, Hamilton Lodge School and College & Dr. Joanna Hoskin, Language & Communication Science Division City, University of London

09:20 GMT
Supporting deaf and hard-of-hearing children to access their linguistic and cultural heritage, Kathryn Margaret Crowe, University of Iceland / Charles Sturt University (Australia) / National Technical Institute for the Deaf (United States of America)

09:40 GMT
On the equal chances of minority populations in all EU countries, Dr. Maria Magdolna Tatár, University of Oslo, retired

10:00 GMT
Translation, adaptation, and imitation, Steve Rushton, UK

10:20 GMT
Kazakh Literature, Bakhtygul Makhanbetova, International literary agent

10:40 GMT
Closing, Senior Lecturer Dr. Alan Libert, University of Newcastle, Australia

Language Shift Reversal Thanks to COVID-19

Valeria Buonocore

Independent Scholar, Newcastle, Australia

vale1stella@gmail.com

Language Shift Reversal Thanks to COVID-19

Abstract: The Italian linguistic situation boasts a notable number of language varieties. Although the data collected by the 2006 ISTAT survey on language use showed that some Italian dialects and other language minorities are suffering from endangerment due to the continuous shift to standard Italian, there are some language varieties, such as Neapolitan, whose language vitality seem to slow down this process. This is due to different factors, such as the high number of speakers, the language prestige in the society, the vast literature, the trends in music and social media, the social interactions and the intergenerational transmission. Particularly, the latters have acquired a greater role in the debate about language endangerment since the beginning of the COVID-19 pandemic in 2020. As a result, scholars are currently researching about its effects on the language in use. Due to the social preventive measures imposed by governments, interactions between people all over the world have changed. Home confinement, social distancing and isolation have affected people's lifestyle and behavior, and therefore their way of communicating. In 2020 spoken interactions have been mostly limited to the inhabitants of the same household or neighborhood, education has been suspended to stop the spread of the virus. Similarly, International, and to some extent, national mobility has been halted; hence, speakers have not felt the need to use the standard variety given the contactless confinement in a domestic environment. It is known that topicality reinforces the use of dialects. It is in this scenario that dialects decline over the last century is likely be put on hold thanks to the COVID-19 pandemic. Moreover, it is possible to assume that syntactic structures, only present in some language varieties, could result into the standard variety, since it is believed that native speakers have the power to contribute to their language development. This could be the case of the differential object marking, only present in some Southern Italian varieties and in other romance languages. The aim of this paper is to investigate the effects of COVID-19 on the language in use and provide the readers with critical guidance regarding future language developments and prospects.

Keywords: Italian varieties, language endangerment and revitalization, language shift reversal

1. Introduction

Numerous languages are at risk of extinction worldwide. "Every two weeks a language dies off in the world" (Crystal 2000;19).

The cultural heritage of mankind is partly represented by local languages and according to the UN's cultural agency, dialects, especially, could soon disappear.

Multilingual realities have long been ignored or rejected. In many countries around the world, from the second half of the twentieth century, the progress of language studies and different sociopolitical and economic phenomena have resulted in an increasing interest in multilingualism. Among the European countries, Italy is one of the biggest representatives of multilingualism. Besides the official language and other minority languages, there are many regional languages which coexist with the standard variety. This situation of diglossia makes the majority of Italians native bilinguals of Italian and at least one other regional variety. These regional varieties differ from each other and from Standard Italian.

In the world, there are roughly 6,500 spoken languages nowadays. However, about 2,000 of those languages have fewer than 1,000 speakers and COVID-19 represents an even bigger threat to their endangerment. As a result, linguists are worried about the possibility of an even further level of endangerment for those local varieties. Given the fact that local and minority languages are transmitted from generation to generation, the advent of COVID-19 has impacted the number of speakers worldwide and older person preyed by COVID-19 might not be able to hand to next generation their native language varieties. Contrastively, home-confinement has contributed to an increase in the use of the local variety. Could this be perhaps beneficial for local languages?

Therefore, the first part of this paper briefly introduces the multilingual Italian territory: from the ´questione della lingua´ to the present times.

Moreover, taking into account the global emergency caused by COVID-19 in 2020, people around the world have spent most of their time in lockdown, i.e. at home. Without minimizing its negative effects, it is possible to find some positive effects in the local language preservations. From a linguistic point of view, the domestic environment promotes the use of the local, minority and/or heritage languages. As a result, the second part of this paper focuses on the effects of the COVID-19 pandemic on the language in use with main focus on the Italian varieties.

After having considered the benefits of COVID-19 for the language in use, the last part of this paper examines the potentiality of a language shift reversal from the standard variety towards the local ones. Giving the example of the differential object marking, a syntactic phenomenon present in some Southern Italian dialects which could result into the Italian standard variety. The aim of this paper is providing a starting point for further investigation in the language development.

2. The multilingual Italian context

According to the UNESCO Atlas of the World's Languages in Danger, there are more than 30 languages in danger in Italy presently.

This unique linguistic situation of multilingualism in Italy is rooted in its historical background. Starting with the *questione della lingua* in the Middle Age (see, e.g., Lepschy 1993: 16–35, and Moss 2000), regarding the debate around which variety among the copious Italian dialects should become the national language, and the affirmation of the Florentine from the cultured class in the Sixteenth century as the most prestigious variety (cf. Maiden 2014), the Italian territory has maintained its status of multilingual country.

Moreover, the coexistence of Italian and minority languages in Italy can be addressed to its late unification, only achieved at the end of the Nineteenth century (cf. Maiden 2014). As a result, the lack of unity has long prevented from having a single official language which was finally achieved in the last decades under the impetus of language policy influenced by demographic, economic and, in part, educational factors.

According to the first ISTAT survey on the use of dialects in 1987/1988, 56.9% of the Italian population was mastering both Italian and the local dialect within the family and preferring mainly the local dialect for spoken interactions.

Although education and language policies resulted in having more than 90% of the people adopting Italian for their interactions, most of them preserves the use of one of the many characteristic dialects of the country or one of the 14 minority languages.

It is common to associate the term *dialects* with "subdivisions of a particular language" (Chambers and Trudgill 1998, 3) or "a system of signs deriving from a common language, living or defunct, normally with a concrete geographical limitation, but not strongly differentiated vis-à-vis others of the same origin"(Alvar 1996, 13). Therefore, it is possible to state that these definitions do not apply to the Italian context, in which dialects have their own system and vitality. According to Tosi (2004), this unusual condition of linguistic diversity in Italy is the result of different factors.

"One factor is that the so-called 'dialects' of Italy are actually Romance languages and not dialects of Italian different from the standard. Another is that Italian is a far less standardized language than other Romance languages" (Tosi 2004, p.1)

Nonetheless, the development of Italian in the past 150 years has been the cause of a language shift from local languages towards Standard Italian. According to Berruto´s calculations, only a little part of the population would still use their dialect at the end of the Twenty-first century (cf. Holtus and Radtke 1994).

Furthermore, the analysis about the major evaluative factors of language vitality drawn up by UNESCO and the data offered by the 2006 ISTAT survey on language use revealed that so-called 'Italian dialects' were suffering a higher exposure to endangerment.

Many external phenomena have been designated by experts as threat to those dialect, such as migration and rapid urbanization, progress, military and/or cultural dominations, but also some fundamental internal forces, such as fashion and trends resulting into younger generations sacrificing the use of 'old-fashioned' dialects in favour of the 'new' standard variety.

On the other hand, some new trends in music and social media have brought the appreciation for identity and 'vintage' again into the limelight. In fact, the demise for some Italian dialects seems to be still far away. Local varieties like Neapolitan, Sicilian and Venetian are still spoken by a large percentage of the population (cf. Coluzzi 2009, 39-54). This is thanks to their variety prestige in the society, their vast literature, the trends in music and social media, the social interactions and the intergenerational transmission. As mentioned above, particularly the latters have acquired a greater role in the debate about language endangerment since the beginning of the COVID-19 pandemic in 2020.

Although the debate in Italy on the vitality of dialects and their future, little has been done to develop strategies to reverse the worrying language shift that both minority languages and dialects are undergoing (cf. Coluzzi 2009, 39-54). Could be COVID-19 the unusual creator of a potential unexpected language shift reversal?

3. Can a virus be beneficial for the preservation of dialects?

2020 has been the year of the coronavirus which has spread across countries worldwide. Its outcome is not only a global health crisis, but also severe economic and socio-psychological consequences.

Due to a range of social preventive measures imposed by the governments, interactions between people all over the world have changed. Home confinement, social distancing and isolation have been affecting people's lifestyle and behaviour, and therefore their way of communicating.

The consequent closure of schools and public amenities, home-office work, forced lockdowns and isolation, the chase for the patient 0, and limited civil liberties have inevitably generated different responses in individuals.

Nonetheless, in the same way in which the world has stopped, language endangerment could be reduced. Isolation and home confinement promote the use of the domestic register which allows minority languages and dialects to acquire new vitality.

With regard to the linguistic situation, the Italian heritage is one of the richest and varied in Europe. If in the past centuries, the deficits in education in Italy have hindered the convergence towards a single language, the current discontinuous formal learning is likely to result into a language shift reversal towards dialects. As a matter of fact, it is known that topicality reinforces the use of dialects; hence, dialects' decline over the last century is likely to be put on hold thanks to the COVID-19 pandemic.

Moreover, Furthermore, the Italian central government has granted more autonomy to the regions in order to have more efficient measures to stop the spread of the virus. The citizens' life is regionally centered and geographically limited. From a linguistic point of view, this has resulted in an increase in the use of the regional varieties.

According to Wicherkiewicz (2001), regional languages tend to undergo different sociolinguistic processes:

"They are used actively predominantly in the cultural life of the region [..] the most common situation being diglossia; monolinguals of regional languages can no longer be found; the education in, and teaching of, regional languages is still hardly developed, [which] deepens [further] the generation gap in active language use; the regional languages are present in the mass-media only at a (sub)regional level: local press, regional radio broadcasting stations, hardly any televisions; religious services in regional languages are still rare; literature in these languages covers only [few] literary domains: mainly regional folklore and traditions, children's literature, poetry, etc. ' In this case, too, most or all of these processes characterize the so-called Italian 'dialects'" (Wicherkiewicz 2001, pp.5-6)

Not to be confused with a regional variety of Italian, or 'regional Italian', which is a variety of Italian showing some traits, particularly in phonology and lexicon, that derive mostly from the local 'dialect'. Although Standard Italian spread through Italy in the 20th century, regional areas of Italy used variations of Italian languages and dialects and became known as Regional Italian (italiano regionale).

From March 2020, people all over the world have been home-confined. Although mass media overwhelmingly use the standard variety, home conversations are the ones which occur the most and now with an even higher frequency. Therefore, dialects and regional Italian varieties have been given a new vitality.

Could this not be a trigger for having some forms of the local varieties affecting the standard and not vice versa, as it was the case before the pandemic?

As a matter of fact, it is possible to assume that syntactic structures, only present in some local varieties, could emerge in the standard variety. It is commonly believed that native speakers have the power to contribute to their language development; hence, speakers of the main local varieties could start using unconsciously some synctactic constructions also in the standard variety.

This could be the case of the 'differential object marking', a syntactic phenomenon present in some romance languages and some Southern Italian varieties but not in Standard Italian.

4. What is the 'differential object marking' and why is this likely to emerge in Standard Italian?

Whenever a language overtly case-marks "some direct objects, but not others, depending on semantic and pragmatic features of the direct object[such as] animacy, defiteness and topicality" (Aissen 2003:435), this can be defined as 'differential object marking' ((henceforth DOM, term coined by Bossong 1985,1991). Animacy, definitess and topicality are mainly arranged along implicational scales (cf. Comrie 1975, Aissen 2003, Croft 2003). According to Aissen (2003, 437), the animacy scale is the main factor for the distribution of the differential object marker. Furthermore, some additional factors such as co-argument asymmetries temporal and aspectual and modal verbal categories have been claimed to be responsible in the distribution of DOM in many languages (cf. Primus 2011; Malchukov and de Hoop 2011).

In Romanian the general differential object marker is the preposition *pe*, whereas in Spanish it is the preposition *a*. as shown in (1):

(1) a. rom. L-am văzut pe Mario. [+HUMAN]
        CL.3.SG.ACC-have.1.SG seen PE Mario.
        'I saw Mario.'
    b. sp. Yo ví a *Mario*.
        1.SG.NOM saw A Mario.
        'I saw Mario.'

Nonetheless, this phenomenon is not present in the standard Italian variety but it is abundantly in use in some Southern Italian varieties such as Neapolitan, Sicilian, Calabrian and Apulian. In these local varieties the prepositional marker a coincides with the prepositional introducer of datives a, as shown in (2):

(2) a. it.   Ho visto Ø Mario.
              have.1.SG seen Ø Mario.
              'I saw Mario.'
       neap. Verette a Mario.
       sicil. Visti a Mario.
       calab. Vitti a Mario.
       apul. Je vist a Mario.

The DOM and its distribution in the Southern Italian varieties has been the topic of extensive research. The analysis in this paper is limited because the material and the arguments presented aim to be a starting point for a broader set of research themes and approaches. For example, the possibility of the prepositional accusative of some Romance languages and dialects to enter the Italian standard variety in the future. This hypothesis is considered in relation to the new reality everyone has had to face during the COVID-19 pandemic. The community's compliance with preventive measure is likely to affect not only its behaviors but also its languages.

Finally, it is possible to believe that in the same way in which COVID-19 has limited the community's freedom, it might slow down (if not reverse) the language shift towards the Standard.

References

Aissen, J., 2003, Differential Object Marking: Iconicity vs. Economy, In: *Natural Language and Linguistic Theory*, (21), 435-483.

Berruto, G. (1994a). Scenari sociolinguistici per l'Italia del Duemila. In: ed. Holtus, G. and Radtke, E.. *Sprachprognostik und das 'italiano di* domani': prospettive per una linguistica'prognostica (Vol. 384), Tübingen: Gunter Narr Verlag.

Berruto, G. (2006). Quale dialetto dialetto per l'Italia del Duemila? Aspetti dell'italianizzazione e risorgenze dialettali in Piemonte (e altrove). In: *Lingua e dialetto nell'Italia del duemila* , ed. Sobrero, A. and Miglietta, A.. Galatina: Congedo Editore.

Bossong, G.,1991, "Differential object marking in Romance and beyond", in D. Wanner and D. Kibbee (eds.,) *New Analyses in Romance Linguistics* (XVIII Linguistic Symposium on Romance Languages 1988), Amsterdam: Benjamins, 143–170.

Chambers, J. K., and Trudgill, P. (1998). *Dialectology*. Cambridge: Cambridge University Press.

Coluzzi, P. (2009). Endangered minority and regional languages ('dialects') in Italy. In: *Modern Italy*, 14(1), 39-54.

Crystal, D. (2000). *Language death*. Cambridge: Cambridge University Press.

D'Alessandro, R. (2016). *When you have too many features: auxiliaries, agreement and DOM in southern Italian varieties*. [online] Available at: http://ling. auf. net/lingbuzz/002908 [Accessed 20 Dec. 2020].

Eberhard, D. M., Simons, G. F., Fennig, C. D. (2019). *Ethnologue: Languages of the world*. SIL International.

Evans, N. (2009). *Dying words: Endangered languages and what they have to tell us* (Vol. 6). New Jersey: John Wiley & Sons.

Hagège, C. (2009). *On the death and life of languages*. Yale University Press.

Maiden, M. (2014). *A linguistic history of Italian*. London: Routledge.

Moseley, Christopher (2010). Atlas of the World's Languages in Danger, 3rd edn. Paris, UNESCO Publishing. Online version available at: http://www.unesco.org/culture/en/endangeredlanguages/atlas [Accessed 17 Nov. 2020].

Onea, E., Hole, D. (2017). Differential object marking of human definite direct objects in Romanian. In: *Revue roumaine de linguistique*, 62(4), 359-376.

Tosi, A. (2004). The language situation in Italy. In: *Current Issues in Language Planning*, 5(3), 247-335.

von Heusinger, K., Kaiser, G. (2005) "The evolution of differential object marking in Spanish", in: K. von Heusinger, G.. Kaiser & E.Stark (eds), *Proceeding of the Workshop Specificity and the Evolution / Emergence of Nominal Determination Systems in Romance*. Konstanz: Fachbereich Sprachwissenschaft der Uni Konstanz, 33-69.

Wicherkiewicz, T. (2001). Becoming a regional language-a method in language status planning? In: *Actes del 2n Congrés Europeu sobre Planificació Lingüística*. Andorra La Vella: Departament de Cultura.

Wurm, S. A. (1991). Language death and disappearance: Causes and circumstances. In: *Diogenes*, 39(153), 1-18.

Bilingual Language Behaviours in Multicultural Australia: An Insight into the Nature of Bilingualism and Directions for Sustainable Community Language Practices

Dr. Iryna Khodos

University of Newcastle, Newcastle, Australia

iryna.khodos@newcastle.edu.au

Bilingual Language Behaviours in Multicultural Australia: An Insight into the Nature of Bilingualism and Directions for Sustainable Community Language Practices

Abstract: In today's pluralistic world, bilingualism is a growing reality, which has furthermore been suggested to contribute to the social and cognitive health of people. Given the inter- and intra-individual variability in bilingual experience, investigating variations in bilingual language practices and further identifying which of them are more efficient in sustaining languages and potentially improving cognitive health of the communities across the world are an important research priority. In the present study, we considered such aspects of bilingual experience as typological proximity/distance, onset age of active bilingualism, language proficiency and language entropy, and explored the way they interplay with (meta)linguistic skills in bilingual adults. Using a background questionnaire and a sentence-judgement task, demographic and language data were collected from 60 linguistically diverse bilingual adults residing in Australia. The results of multiple regression analyses revealed that three of the language variables considered – language proficiency, typological proximity/distance and onset age of active bilingualism – accounted for the variance in metalinguistic data. Specifically, higher levels of language proficiency, use of typologically closer languages and earlier onset age were related to higher metalinguistic scores. These findings reinforce the view on bilingualism as a multidimensional experience, whose consequences depend on a number of distinct but interrelated language learning and use variables. Taken together with a high correlation between language use and language proficiency, the results also suggest that bilinguals who have equally used two languages in the same contexts but with different speakers are more likely to obtain and maintain higher levels of proficiency in both languages, which together with an earlier active use of two typologically closer languages may allow bilinguals to experience advantages in metalinguistic skills. These findings, therefore, underscore the importance of educational programmes and community language practices, which allow people to learn and equally use each of their languages, for maintaining and further developing two/multiple languages.

Keywords: bilingual experience, metalinguistic awareness, language proficiency, language behaviours

1. Introduction

The ability to express the same thought in two different languages is considered to lead to an increased awareness of formal and substantive properties of language (i.e. enhanced metalinguistic awareness; Bialystok, 2001; Galambos and Goldin-Meadow, 1990; Jessner, 2008; Lambert, 1990; Vygotsky, 1962). Given that metalinguistic advantages may, in turn, generalise to other areas of cognitive abilities and thereby contribute to cognitive health, investigating the mechanism underlying metalinguistic benefits in bilinguals is an urgent research task. This is especially true in multicultural Australia, where one in five people speaks a non-English language in addition to English (Australian Bureau of Statistics, 2016). Exploring (meta)linguistic dimensions of bilingualism in this context is, therefore, also crucial for understanding matters concerning integration and socialisation of people as well as for political and educational decision-making. In the present study, we extended the previous metalinguistic research on bilingualism by exploring the interplay between specific dimensions of language experience and metalinguistic skills in bilingual adults in multicultural Australia.

The idea of bilingualism boosting metalinguistic awareness was first expressed by Vygotsky (1962). The psychologist was, moreover, the first to point to the possibility of metalinguistic advantages to contribute to cognitive benefits across the cognitive domains, with the effects depending largely on the metalinguistic skills induced by the use of more than one language. To investigate this enticing hypothesis, Bialystok and Ryan (1985) suggested and innovatively implemented the dual component model. Conceptualising metalinguistic awareness as a form of language processing, the researchers developed the tasks targeting its two skill components: (1) the analysis of linguistic knowledge into structured categories, and (2) the control of attentional procedures to select and process specific linguistic information. For instance, on the basis of existing word awareness tests, Bialystok et al. (2003) developed a sound-meaning task, which required participants to select which of two words matched a target for either the sound (rhyme) or meaning (synonym). Also, Bialystok (1986) manipulated the characteristics of a sentence-judgement task by constructing sentences that were grammatically correct, grammatically incorrect but meaningful or semantically anomalous but grammatical.

The results of Bialystok's empirical investigations showed that bilingualism does not have a direct effect on metalinguistic awareness rather it influences the two underlying skill components and, what is more, in a different way. In her studies, the bilingual advantage was seen primarily in tasks demanding a high level of control of linguistic processing. In these tasks, the specific skills of the participants in L1 and L2 were not shown to affect their performance. Thus, a superior performance of bilinguals was seen to be due to the early bilingual experience of dual language systems and frequent attention to formal aspects of language. On the other hand, in tasks requiring high levels of analysis, findings were somewhat mixed and depended on the combined proficiency of bilinguals in both languages.

Bilinguals were shown to outperform monolinguals only if they had high levels of proficiency.

Bialystok's findings are in line with a number of other studies on metalinguistic awareness (Davidson et al., 2010; Galambos and Goldin-Meadow, 1990; Hakuta and Diaz, 1985; Ricciardelli, 1992). Similar to Bialystok, these researchers point to a positive influence of mastering two languages on person's ability to control the processing of linguistic information. These findings are also consistent with the 'threshold hypothesis' proposed by Cummins (1977), according to which an overall bilingual superiority in terms of metalinguistic abilities is found only for those who have attained a high degree of bilingualism.

Given the multidimensional nature of bilingualism (de Bruin, 2019; Khodos and Moskovsky, 2020; Laine and Lehtonen, 2018), we expected metalinguistic skills to be affected by different language experiences differently. However, most of the previous metalinguistic studies have not considered the inter-individual variability in their participants' language experience. Instead, they treated bilingualism and monolingualism as categorical constructs and compared bilinguals and monolinguals as two distinct groups, with their members categorised either as a homogenous whole or in terms of binary oppositions (e.g., early vs late, simultaneous vs sequential, more proficient vs less proficient, L1-dominant vs L2-dominant, balanced vs unbalanced). Therefore, they were not able to uncover the mechanism underlying metalinguistic advantages in bilinguals.

In order to address the limitations of previous metalinguistic research, we examined metalinguistic skills in linguistically diverse bilingual adults, with the inter-individual variability in their bilingual experience being analytically taken into account. Specifically, we explored the following research questions: 1) which combination of bilingual experience accounts for the most variance in bilinguals' metalinguistic performance, and (2) the extent to which each variable in the combination contributes to explaining the variance in participants' metalinguistic scores. Given the peculiarity of the bilingual sample in the present study, we were able to consider the following dimensions of bilingual experience: (1) typological proximity/distance between two languages (Germanic languages/smaller distance and non-Germanic languages/larger distance), (2) onset age of active bilingualism (age at which they began using both languages actively on a regular basis), language proficiency (average language proficiency in two languages) and language entropy (the frequency of use of two language).

2. Procedure

Bilingual adults (aged 20-40 years) speaking English as a second language were recruited from the research sites located in the Newcastle/Hunter area, NSW, Australia. The participants were asked to fill in the Language and Social Background Questionnaire

(Anderson et al., 2018) to get data on key demographic and language variables. Following that, the bilinguals completed the sentence-judgement task (Bialystok and Barac, 2012; Moreno et al., 2020) for their metalinguistic skills to be assessed.

2.1 Study setting

All the participants resided in Australia, which has a heterogeneous population compared with other countries in the world. There are Indigenous people, descendants of the original UK settlers and a diverse group of immigrants, who either come to Australia as bilinguals or develop bilingual knowledge in the years following their arrival. Considering this, Australia is a multilingual and multicultural country, with the official language, English, coexisting with Aboriginal and immigrant languages.

According to the 2016 Census data (Australian Bureau of Statistics, 2016), one in five Australians now speaks a language other than English at home. Among them, the most commonly spoken ones are Mandarin, Arabic, Cantonese, Vietnamese and Italian, as compared to Italian, Greek, Cantonese, Arabic, Mandarin and Vietnamese in 2006. This points to two opposite tendencies in Australian society: a substantial decrease in the home use of a number of European languages (in particular German, Italian and Greek) and a great increase in Asian languages, especially Mandarin. In other words, the linguistic diversity of Australia is shifting away from the European languages of the post-war period to languages of new migration waves, mainly from Asia and the Middle East (Clyne et al., 2008).

The substantial number of non-English languages notwithstanding, Australia remains a strongly Anglocentric country, where the dominance of English is largely unchallenged (Rubino, 2010). As pointed out by Clyne (2005), the 'monolingual mindset' is still one of the key challenges of modern Australian society. The majority of its native English speakers do not speak any other language. Moreover, they show limited interest in languages and/or language study. In addition to cultural and social attitudes, the limited availability and accessibility of language programs in Australian institutions is another possible barrier to cultivating Australian bilingualism/multilingualism (Rubino, 2010).

In this light, bilingualism and multilingualism in Australia appear to be represented mainly by Aboriginal people and immigrants from non-English speaking backgrounds (individual bilingualism). However, even they tend to abandon their native languages relatively quickly as a consequence of lack of opportunities to apply their native language in broader social contexts (mostly single-language contexts) and lack of institutional support.

2.2 Participants

The sample consisted of 60 bilingual adults (20-40 years old), including 22 males and 38 females. All of them held a higher university degree – either Bachelor's or Master's degree (M = 4.00, SD = .00). Thus, education was not considered in further analysis. Descriptive statistics are provided in Table 1.

The bilinguals were from varied non-English speaking backgrounds. Their first language belonged to one of the following language branches: Germanic (11); Romanic (13); Slavic (7); Iranic (9); Indo-Aryan (5); Sinic and Tibeto-Burman (6) and other (9), including Vietnamese (2), Greek (1), Cambodian (1), Azerbaijani (1), Malay (1), Filipino (1), Malayalam (1) and Shona (1). Most of them had started acquiring English in childhood (M = 9.35, SD = 4.64) in a single first language-oriented environment and had begun using both languages on a regular basis (in the same or different contexts) shortly before or upon arriving in Australia (i.e. onset age of active bilingualism; M = 21.33, SD = 7.83).

Table 1: Descriptive statistics for demographic and language variables

Variables	N	Mean	SD
Demographic variables			
Gender	male – 22 female – 38	-	-
Age	60	31.92	4.45
Education	60	4.00	.00
Language variables			
Onset age of active bilingualism	60	21.33	7.83
L1 [non-English] proficiency	60	9.33	.76
L2 [English] proficiency	60	8.16	1.00
Language use in close social context	60	3.53	.50
Language use in broad social context	60	1.98	.13

Note. Age and onset age in years. Education on a 4-point scale (1 = upper secondary, 2 = post-secondary non-tertiary, 3 = short-cycle tertiary, 4 = tertiary education). Language proficiency on a 10-point scale (0 = no proficiency, 10 = high proficiency). Language use on a 5-point scale (1 = all English; 3 = half English, half the other language; 5 = only the other language).

In Australia, the bilinguals were immersed in a mostly single second language-oriented environment: on average, they indicated the use of mostly English in terms of broad social contexts (M = 1.98 on a 5-point Likert scale, SD = .13). Nevertheless, the participants varied in the way and extent to which each of the two languages was used in close social contexts

(M = 3.53, SD = .50) and language proficiency (L1 [non-English] proficiency: M = 9.33 on a 10-point Likert scale, SD = .76; L2 [English] proficiency: M = 8.16, SD = 1.00).

2.3 Instruments

The Language and Social Background Questionnaire was based on the well-established research tool by Anderson et al. (2018). In line with the original questionnaire, the one used for the purposes of the current study consisted of three sections. The Social Background Section captured demographic information, including age, gender, highest level of education, immigration status. The Language Background Section assessed the number of languages spoken and proficiency for speaking, listening, reading and writing the indicated language(s). Finally, the Community Language Use Behaviour Section elicited information on the language usage at different life stages and in different social contexts.

The sentence-judgement task was developed in accordance with the cross-validated dual component model of metalinguistic awareness (Bialystok and Ryan, 1985). It consisted of 24 sentences presented in context, as part of three short passages. In line with the previous studies using this type of task (e.g., Bialystok and Barac, 2012; Moreno et al., 2020), the sentences were constructed along two linguistic dimensions: a semantic one and a grammatical one. This resulted in four sentence frames: grammatical, meaningful (GM; 6 items), grammatical anomalous (Gm; 6 items), ungrammatical, meaningful (gM; 6 items) and ungrammatical anomalous (gm; 6 items). This way it was possible to target analysis and control components separately: the highest level of analysis was required to deal with gM sentences, while the highest level of control was needed to judge Gm sentences (Bialystok, 2001).The participants were given 20 minutes and were asked to judge whether the given sentences were grammatical or ungrammatical irrespective of their meaning. The key point was that judgements had to be made on the basis of how each of the sentences was used in the given text. In case the ungrammatical option was selected, correction was required.

3. Results

The obtained background and metalinguistic data were subjected to multiple regression analyses conducted in R (version 3.6.1). The participants' sentence-judgement task scores were used as dependent variables. The 24 items were combined according to the sentence frame (GM, Gm, gM and gm), which resulted in four factors consisting of six relevant components. Descriptive statistics are given in Table 2.

Table 2: Descriptive statistics for the dependent variables

Dependent variables	N	Mean	SD
GM	60	3.99	1.27
Gm	60	4.00	1.40
gM	60	3.45	1.37
Gm	60	3.57	1.50

Note. Number of correct sentence-judgement task items out of six.

Demographic and language variables were entered as predictors. In particular, each regression included two demographic variables: *gender* (1 = male, 0 = female) and *age* in years; and a set of language variables: *typological proximity/distance* between two languages, *onset age of active bilingualism, language proficiency* and *language entropy*.

Typological proximity/distance was extracted from the data on bilinguals' L1 and used as a dummy variable: 1 = Germanic languages, 0 = non-Germanic languages. *Onset age of active bilingualism* was based on the age at which the bilinguals began using their two languages actively on a daily basis and included as a continuous variable in years. *Language proficiency* was computed on the basis of the average proficiency score for each language by using the calculation as in Vaughn and Hernandez (2018):

$$(L1+L2) \times \sqrt{\frac{2 \times L1 \times L2}{L1^2 + L2^2}} \qquad (1)$$

Language proficiency was treated as a continuous variable (0 = no proficiency in each of the languages, 20 = high proficiency in both languages). *Language entropy*, i.e. a continuous measure of how often languages are used (0 = only one language is used, 1 = each language is used equally), was calculated using the equation as in Gullifer et al. (2018):

$$H = -\sum_{i=1}^{n} P_i \log_2(P_i) \qquad (2)$$

Here, *n* represents the total possible languages (two in the present study) and P_i represents the proportion associated with the use of a given language. The proportion of L1 and L2 use for each bilingual was quantified on the basis of the self-reported language use data. Means and standard deviations for the predictors are provided in the Table 3.

Table 3: Descriptive statistics for the predictors

Predictors	N	Mean	SD
Demographic			
Gender	male – 22 female – 38	–	–
Age	60	31.92	4.45
Language			
Typological proximity/distance	Germanic – 11 non-Germanic – 49		
Onset age of active bilingualism	60	21.33	7.83
Language proficiency	60	17.50	1.46
Language entropy	60	.64	.35

Note. Gender: 1 = male, 0 = female. Age and onset age in years. Typological proximity/distance: 1 = Germanic languages, 0 = non-Germanic languages. Language proficiency on a 20-point scale (0 = no proficiency in each of the languages, 20 = high proficiency in both languages). Language entropy on a 1-point scale (0 = only one language is used, 1 = each language is used equally).

Given a statistically significant correlation between language proficiency and language entropy, $r = .50$, $p < .001$, we created two base-line models. Both contained *gender, age, typological proximity/distance* and *onset age of active bilingualism*. However, one had *language proficiency* and the other included *language entropy*. Following that, we performed multiple regressions with backward elimination using the regsubsets function. The results of the analyses showed that the best-fitting model among all for the sentence-judgement task items was the one with *language proficiency* among the predictors (see Table 4).

Table 4: The best-fitting models showing the capacity of language variables to predict the sentence-judgement task items

Variables	B	SE*B*	t	Sig.
Gm: $R^2 = 37.5\%$, $p < .001$				
$\Delta R^2 = 37.6\%$, $p < .001$				
Typological proximity/distance	-1.52	.39	-3.86	.001
Onset age of active bilingualism	-.08	.02	-2.64	.05
Language proficiency	.45	.14	3.13	.01
gM: $R^2 = 29.3\%$, $p < .001$				
$\Delta R^2 = 26.8\%$, $p < .001$				
Typological proximity/distance	-.71	.39	-1.81	.05
Language proficiency	.62	.14	4.29	.001
gm: $R^2 = 15.8\%$, $p < .001$				
$\Delta R^2 = 14.3\%$, $p < .001$				
Language proficiency	.56	.17	3.3	.001

Note. Onset age of active bilingualism in years. Typological proximity/distance: 1 = Germanic languages, 0 = non-Germanic languages. Language proficiency on a 20-point scale (0 = no proficiency in each of the languages, 20 = high proficiency in both languages).

In the case of gm, *language proficiency* was the only predictor in the model ($R^2 = 15.8\%$, $p < .001$). As the bilinguals' language proficiency increased by one point on a 10-point scale, their gm scores increased by .56 points. In the case of gM and Gm, the best model contained more than one predictor. In addition to *language proficiency*, the best model for gM included *typological proximity/distance* ($R^2 = 29.3\%$, $p < .001$); whereas the best model for Gm contained *language proficiency, typological proximity/distance* and *onset age of active bilingualism* ($R^2 = 37.5\%$, $p < .001$). In both cases, the participants whose L1 belonged to the Germanic language family performed better (gM: $B = .71$, $p < .05$; Gm: $B = 1.52$, $p < .001$). The bilinguals also obtained higher scores as their language proficiency increased by one point on a 10-point scale (gM: $B = .62$, $p < .001$; Gm: $B = .45$, $p < .01$). In the case of Gm, the participants' scores furthermore increased by .08 points as their onset age of active bilingualism decreased by one year. As for the GM items, none of the models explained the variance in the scores, $p > .05$.

4. Discussion and conclusion

The study tested linguistically diverse bilingual adults residing in Australia on the sentence-judgement task measuring two metalinguistic skills, i.e. the analysis of linguistic knowledge and the control of attentional procedures. This was done to investigate: (1) which combination of bilingual experience (if any) – *typological proximity/distance, onset age of active bilingualism, language proficiency* and *language entropy* – accounts for the variance in bilinguals' metalinguistic performance, and 2) the extent to which each variable in the combination contributes to explaining the variance in the participants' metalinguistic skills.

The results of the study showed that variance in participants' metalinguistic performance was related to differences in their bilingual experience. Specifically, the model with with *language proficiency* among the predictors accounted for the most variance in metalinguistic scores. According to the data, language proficiency was predictive of the participants' performance on all the sentence-judgement task items: higher levels of language proficiency were related to higher scores. The use of two typologically closer languages further contributed to better performance on the task items requiring the highest level of analysis (gM), and together with an earlier onset of active bilingualism, it was also related to higher scores on the task items placing the greatest burden on control (Gm).

Consistent with recent studies, our research suggests that particular dimensions of bilingual experience rather than bilingualism per se are linked to bilingual advantages (Bialystok and Barac, 2012; Gullifer et al., 2018; Khodos and Moskovsky, 2020). However, our study is unique in that it provides insight into dimensions of bilingual experience which may boost and further maintain enhanced (meta)linguistic skills in adults. In particular, the results of the multiple regression and correlation analyses suggest that bilinguals who have equally used two languages in the same contexts but with different speakers are more likely to obtain/maintain higher levels of proficiency in both languages, which, in turn, may be related to enhanced analysis skills. When combined with the use of two typologically closer languages and an earlier onset of active bilingualism, higher language proficiency may furthermore allow bilinguals to experience advantages in control skills.

Given that most of the previous studies have not considered the inter-individual variability in bilingual experience while interpreting their participants' metalinguistic performance, it is quite difficult to reconcile the present findings with the wider literature on metalinguistic awareness. Among the variables explored in the present study, the role of the typological proximity/distance between L1 and L2 appears to have received the least attention. This might stem from the fact that most previous research has worked with participants that were linguistically homogeneous – same L1 and same L2. However, even when the participants varied in their language pairs, the researchers did not control for the typological proximity/distance variable in their studies. Our findings, therefore, extend the previous metalinguistic studies and call for a need to focus more research attention on the individual features of bilingual experience and the ways they interplay with language and other cognitive domains (for related ideas, see de Bruin, 2019; Khodos et al., 2020; Laine and Lehtonen, 2018).

The need for further research notwithstanding, the present study contributes clearly to our understanding that bilingual experience can offer advantages that extend beyond language. These benefits, as a result, can have socially relevant consequences for educational attainment and future socioeconomic success. In the context of multicultural Australian, this underscores the importance of introducing suitably designed educational, social and political policies encouraging bi-/multilingualism and creating the best possible setting/environment for learning and using two/multiple languages. Establishing language learning programmes and promoting social practices that maintain and further develop Indigenous and community languages seems particularly desirable. Along other bi-/multilingual practices, this may have serendipitous benefit of improving the social and cognitive health of multicultural Australia.

References

Anderson, J.A., L. Mak, A.K. Chahi and E. Bialystok. (2018). The language and social background questionnaire: Assessing degree of bilingualism in a diverse population. *Behavior Research Methods,* **50**, pp. 250–263.

Australian Bureau of Statistics (2016). *Census Dictionary: Australia 2016*, Catalogue No. 2901.0, ABS, Canberra.

Bialystok, E. (1986). Children's concept of word. *Journal of Psycholinguistic Research,* **15**, pp. 13–32.

Bialystok, E. (2001). Metalinguistic aspects of bilingual processing. *Annual Review of Applied Linguistics,* **21**, pp. 169–181.

Bialystok, E. and R. Barac. (2012). Emerging bilingualism: Dissociating advantages for metalinguistic awareness and executive control. *Cognition,* **122**(1), pp. 67–73.

Bialystok, E., S. Majumder and M. Martin. (2003). Developing phonological awareness: Is there a bilingual advantage? *Applied Psycholinguistics,* **24**, pp. 27–44.

Bialystok, E. and E.B. Ryan. (1985). Toward a definition of metalinguistic skill. *Merrill-Palmer Quarterly,* **31**(3), pp. 229–251.

Clyne, M. (2005). *Australia's language potential*. Sydney: University of New South Wales Press.

Clyne, M., J. Hajek and S. Kipp. (2008). Tale of two multilingual cities in a multilingual continent. *People and Place,* **16**(3), pp. 1–8.

Cummins, J. (1977). Cognitive factors associated with the attainment of intermediate levels of bilingual skills. *Modern Language Journal,* **61**, pp. 3–12.

Davidson, D., V.R. Raschke and J. Pervez. (2010). Syntactic awareness in young monolingual and bilingual (Urdu-English) children. *Cognitive Development,* **25**(2), pp. 166–182.

De Bruin, A. (2019). Not all bilinguals are the same: A call for more detailed assessments and descriptions of bilingual experiences. *Behavioral Sciences,* **9**, pp. 1–13.

Galambos, S.J. and S. Goldin-Meadow (1990). The effects of learning two languages on levels of metalinguistic awareness. *Cognition,* **34**(1), pp. 1–56.

Gullifer, J.W., X.J. Chai, V. Whitford, I. Pivneva, S. Baum, D. Klein and D. Titone. (2018). Bilingual Experience and Resting-State Brain Connectivity: Impacts of L2 Age of Acquisition and Social Diversity of Language Use on Control Networks. *Neuropsychologia,* **117**, pp. 123–134.

Hakuta, K. and R.M. Diaz (1985). The relationship between degree of bilingualism and cognitive ability: A critical discussion and some new longitudinal data In K.E. Nelson (Ed.), *Children's language* (Vol. V, pp. 319–344). Hillsdale, NJ: Lawrence Erlbaum.

Jessner, U. (2008). A DST model of multilingualism and the role of metalinguistic awareness. *Modern Language Journal,* **92**(2), pp. 270–283.

Khodos, I. and C. Moskovsky. (2020). Dimensions of bilingualism promoting cognitive control: Impacts of language context and onset age of active bilingualism on mixing and switching costs. *Linguistic Approaches to Bilingualism.* Advanced online publication. http://doi.org/10.1075/lab.19064.kho

Khodos, I., C. Moskovsky and S. Paolini. (2020). Bilinguals' and monolinguals' performance on a non-verbal cognitive control task: How bilingual language experience contributes to cognitive performance by reducing mixing and switching costs. *International Journal of Bilingualism.* Advanced online publication. https://doi.org/10.1177/1367006920946401.

Lambert, W.E. (1990). Persistent issues in bilingualism. In B. Harley, A. Patrick, J. Cummins, & M. Swain (Eds.), *The development of second language proficiency* (pp. 201–218). Cambridge: Cambridge University Press.

Laine, M. and M. Lehtonen. (2018). Cognitive consequences of bilingualism: where to go from here? *Language, Cognition and Neuroscience,* **33**(9), pp. 1205–1212.

Moreno, S., E. Bialystok, Z. Wodniecka and C. Alain. (2020). Conflict resolution in sentence processing by bilinguals. *Journal of Neurolinguistics,* **23**(6), pp. 564–579.

Ricciardelli, L.A. (1992). Bilingualism and cognitive development in relation to threshhold theory. *Journal of Psycholinguistic Research,* **21**, pp. 301–316.

Rubino, A. (2010). Multilingualism in Australia: Reflections on current and future research trends. *Australian Review of Applied Linguistics,* **33**(2), pp. 17.11–17.21.

Vaughn, K.A. and A.E. Hernandez. (2018). Becoming a balanced, proficient bilingual: Predictions from age of acquisition & genetic background. *Journal of Neurolinguistics,* **46**, pp. 69–77.

Vygotsky, L.S. (1962). *Thought and Language.* Cambridge: M.I.T. Press.

A Critical Examination of Teaching Methodology in ELICOS in Australia: A transition to online course under the COVID-19 crisis

Sarah Mengshan Xu

University of Newcastle, Australia

Mengshan.Xu@uon.edu.au

A Critical Examination of Teaching Methodology in ELICOS in Australia: A transition to online course under the COVID-19 crisis

Abstract: This study was designed to elucidate the immense online pedagogy and the upskilling in language teaching technology in English Language Intensive Courses for Overseas Students (ELICOS) under the COVID-19 pandemic. To discover what teaching methodologies and how they were employed in online courses to replace the traditional teaching pedagogies, classroom observations were conducted in the Language Centre at the University of Newcastle, Australia. The data were collected through Zoom platform over a month from two ELICOS online classrooms, intermediate and upper-intermediate, and students (N = 31) and teachers (N = 2) were observed. The findings indicated some similarities with previous studies that investigated under SARS epidemic, while more up-to-date technology implementations, including Zoom and OneDrive, were used. The eclectic method, namely, Communicative Language Teaching (CLT) and Task-based Language Teaching (TBLT), along with Information and Communication Technology (ICT), was adopted in daily courses. The final discussion focused on some of the problems inherent in online forum and recommendations were made for further research. Some potential values came from this research in the form of encouraging the readers who were not aware of the challenges that might be encountered in virtual classes to be reflexive with their curriculum designs and teaching methodologies.

Keywords: Teaching methodology; ICT; ELICOS; COVID-19

1. Introduction: New Challenges for Teaching and Learning

The COVID-19 outbreak had rapidly hit the world and within several months had a devastating impact on the world's economy and caused changes in the consumption of education. One in five students worldwide was staying away from school due to the COVID-19 crisis while another one in four was barred from higher education institutions (UNESCO, as cited in Owusu-Fordjour, Koomson and Hanson, 2020). The situation in general language education in Australia has changed under this pandemic. With the cooperation of 'social distancing' required by the government and the suspension of face-to-face classes, most universities in Australia and New Zealand have closed but resumed with online teaching. Other school services remain open, but teaching is restricted to delivering courses online exclusively (Moorhouse, 2020).

Numerous Language Centre and ESL classrooms went through an educational transition to the online form of distance teaching and learning, and live communicative platforms like Zoom and Collaborate were giving crucial support under this challenging time. As most of the courses were designed for real-time interactions in physical classrooms, teachers are no doubt facing enormous challenges to adapt their traditional teaching methodologies into distance teaching operations while still regarding the needs of students. It comes to light that the pandemic might also have negative effects on students' learning outcomes, especially for those who fail to adapt themselves to the e-learning environment. Whether this quick transition to distance course, where the language learning environment created by using the internet, video/audio/text communication and software, could be successful is still an unresolved question (Basilaia and Kvavadze, 2020).

2. Related Literature on Teaching Methodologies

In order to recognise the methodologies teachers employ in online classes, there is cause for reviewing relevant historical research and explore the data collected from traditional classrooms. In this paper, three teaching methodologies will be introduced.

2.1 Communicative Language Teaching (CLT)

Due to the absence of phonological instruction in language classes in the previous decades and the importance of learners' unrehearsed responses in real-life communication, CLT has been gradually introduced into classes and has seen widespread use until today. A report by Savignon (1991) noted that introducing systematic communicative materials under various social contexts, including understanding the implication of appropriateness and the cultural norms, were all seen as the way to improve communicative competence and form the central core of the communicative approach. The advantages of the CLT were further identified by Savignon (1991) at the University of Illinois. Students who were experiencing an 18-week communicative teaching program, in which they were encouraged to take the risk to make spontaneous utterances instead of focusing on the grammatical knowledge and memorising patterns, the level of their communicative competence in the final task substantially surpassed other counterparts.

However, this pedagogical methodology has also been criticised for its lack of structure and accuracy. One who is in favour of this meaning-focused teaching approach might achieve fluency but not accuracy in language acquisition. According to Sato and Kleinsasser (1999), some negative concepts, including neglecting grammar, over-emphasizing speaking, and involving time-consuming activities, were frequently associated with CLT. Also, a class observation applied in an Australian state school discovered that the notion of the CLT still remains arguably vague among ESL teachers. Most of the participants confused the substantial similarity between audio-lingual method and the CLT with the same goal of teaching, and the significant overlapping areas in-between CLT and the natural approach due to the sharing of teaching syllabus (Koondhar, Siming and Umrani, 2018; Sato and Kleinsasser, 1999).

2.2 Task-based Language Teaching (TBLT)

In recent decades, TBLT has been considered as one of the most effective pedagogies. According to Dim (2013), tasks could potentially put learners into a more effective, self-directed, and more accessible situation. By working on the tasks, students can track their learning progress as the tasks could be a 'monitor' of their works. And it helps them to open and close conversations, to work together naturally, and to interrupt and confront. Some students believed that TBLT made them more independent thinkers since they were expected to accomplish the tasks individually. Some positive feedback indicated that TBLT-oriented classrooms made it easier for students to remember what they had learned, improving their self-confidence, and made them feel more curious about learning (Tomlinson and Dat, 2004).

However, language teachers who apply the TBLT in the classroom might find it difficult to evaluate the learners' progress if they fail to familiarise themselves with assessment methods of the TBLT (Thi Ly, 2018). Also, receiving feedback from the teacher after the test, as one of the main characteristics of the TBLT, might have some negative effects on students. In the majority of language classrooms, the teachers are encouraged to give valuable feedback, so that learners can realise their weaknesses and make adjustments during the learning process, while it might not work this way for those who are acutely sensitive to the evaluations. According to the interviews in Horwitz and Cope's research (1986), students complained that they were afraid of the teacher correcting every mistake they made, and this statement affirms the questionnaire results collected by Tomlinson and Dat (2004), which indicated that 12.6% of the learners feel apprehensive when receiving teacher's criticism.

2.3 Information and Communication Technology (ICT)

Since the implementation of computers and various technologies in educational systems, ICT has been developed and widely used in the ESL context. The effective use of ICT can transform the traditional teaching and learning process into a creative, self-directed and constructive way (Shah and Empungan, 2015). This teaching methodology is highlighted as effective for language learning by affording access and exposure to authentic materials, communication opportunities, and instant individualised feedback (Røkenes and Krumsvik, 2016). Also, Yunus, Nordin, Saleh, Embi and Salehi (2013) showed that ICT could be beneficial in term of motivating and attracting learners' attention as the lessons were more interesting when involving technological tools.

However, the successful integration of ICT is highly dependent on the preparation and attitudes of instructors. Teachers may find a gap between what they expect in ICT and the practices they encountered during the school practicum. Several researchers have identified that teacher's digital competence is often limited to basic digital skills such as office tools and social media while having little experience with using ICT for delivering class (Yunus et al., 2013). The reason for this weakness was identified as the lack of support in the use of interactive digital tools other than office software in teacher education programs and partnership schools. Educators were often frustrated and frequently mentioned that they have not received any training for how ICT should be used in education, therefore causing challenges for them to adapt ICT effectively and develop digital competences (Røkenes and Krumsvik, 2016). Meanwhile, students found it difficult to develop the technical competence required for the educational practices since they did not have enough knowledge and practical skills including using databases and searching for articles (Tour, 2010). Also, the use of computer technology might lead to a 'lackadaisical attitude' among students whereby they were not taking works seriously (Røkenes and Krumsvik, 2016). The nature of the ICT is more likely to distract students' attention from the teacher's instruction, therefore making class control difficult for the instructors.

3. Research Questions

In response to the need for more research into online teaching methodology replacements to face-to-face instructional approaches, this study aims to address the following key questions:

> (1) What teaching methodologies were employed in online courses in the Language Centre at the University of Newcastle?
>
> (2) How were these pedagogies adapted to the virtual teaching environment?

4. Methodology

4.1 Participants

The participants of the study were 31 students studying in online ELICOS program and 2 teachers who were taking charge of the classes. The initial recruitment process planned to involve 50 students and 4 instructors approximately, although due to the constraints of the COVID-19 the number of the volunteers eventually dropped to 33. The student participants consisted of two different levels, intermediate and upper-intermediate, and were a mix of international students who were not fluent in English. The data for this study were collected from these two groups of students with their teachers for over a month. Also, the selected volunteers were not screened for demographic variables such as age, gender and ethnicity, and only those who formally indicated their consent to participate were recruited.

4.2 Instruments

The observations of two different virtual classes over a month were accomplished to analyse the adaption of teaching methodologies in an online environment. All the classes were observed through the Zoom platform and each of the classes met for 12 hours per week. In order to conduct the observations in a structured and controlled way, the descriptive field notes were taken to describe how the teaching methodologies were adopted. Some elements were captured multiple times, indicating the frequency of the used pedagogy. Meanwhile, the results of the observations were confirmed by the teacher at the end of the class, to corroborate the information collected and to reduce bias and evidence reporting errors. Also, during the process of observing, the role of the researcher in the intermediate course began as a nonparticipant observer and then transformed into a participant observer as advised by both teacher and students. In many observational situations, to be subjectively involved in the setting and to see the phenomenon more objectively, it is beneficial to shift roles when a researcher adapts into the setting (Creswell, 2002).

5. Findings

The research questions of the study aim to find out what teaching methodologies teachers choose to employ in the online ELICOS and how well they were adapted in the online environment. In order to address these objectives, the data were collected through the administration of the semi-structural classroom observations. Both observed classes commenced with a roll call from teachers to check on student's attendance, with a review of previous homework, and ended in the manner of assigning homework for the following class. After all the field notes were examined, the different combinations of multiple teaching methodologies were discovered, and classified into three themes.

5.1 The Combination of TBLT and ICT

Figure 1.

TBLT		ICT	Frequency
- Assigned tasks.	↔	OneDrive; Zoom.	45
- Provided feedback.	↔	Onedrive.	18
- Explained the task instructions and purposes.	↔	Zoom; OneDrive.	25
- Mimicked the exam conditions.	↔	Zoom.	3

Note: frequency = total times that was captured during the observations.

Figure 1 above presents the analysis of the adopted pedagogies in observed classroom practices. In order to simulate the regular classroom conditions, both teacher participants were intensely using 'share screen' function in Zoom classrooms to demonstrate the course PowerPoint, shared pictures, sought information on Google, as well as adopted informative videos from Youtube. Meanwhile, OneDrive, a Microsoft Office product, was another ICT tool frequently used in the observed classes, which allowed teachers and students to collaborate by using 'real-time co-authoring' and other helpful features. As the frequency data show above, a large amount of time was spent on assigning and completing tasks. In detail, learners were expected to answer the questions listed on the PowerPoint and typed it down on the same slide or into the chatbox, according to teacher's preference. All the course materials were able to access from OneDrive with the provided data of how many viewers and views of a certain digital file, therefore providing a better picture for the teacher to monitor learners' studies. Also, in corresponding with the virtual environment, most of the reading materials were scanned and transformed from the traditional textbook to digital files with good graphics quality.

Before starting the tasks, students were required to access the correspondent file, complementing with the question list and task instructions presented on the 'share screen'. The teacher subsequently stated what learners were expected to do with the purposes and benefits of the task. Following rule clarifications, the task continued and students worked independently in the Zoom room and the answers were shared both orally or with the use of the 'real-time co-authoring' feature in OneDrive. Finally, at the end of the lesson, students were assigned homework taken from an authentic text. The teacher would receive an informed email from OneDrive once the student submitted their works. As such, feedback and comments on after-class tasks would be left on the learner's original file through OneDrive. In contrast, for in-class tasks, teachers used body language, such as clapping and giving thumbs up, positive facial expression and verbal complement through Zoom camera and microphone to support learners. Also, during the exam week, it was observed that the teacher played the examiner role to start the task to familiarise students themselves to the online exam condition.

5.2 The Combination of CLT and ICT

Figure 2.

CLT		ICT	Frequency
- Used visual aids (e.g. pictures) and had conversations about it.	↔	'Share screen' in Zoom.	9
- Students produced a short speech or oral presentation.	↔	PowerPoint; 'Share screen'.	12
- Minor corrections when learners speak, big encouragement.	↔	Zoom.	16
- Called on students to read the question instructions.	↔	OneDrive; Zoom.	13

The usage of CLT and ICT methodologies were illustrated in Figure 2. Both intermediate and upper intermediate classes proceeded similarly throughout the whole observation period. Most frequently, students were told to read the question's instructions through 'share screen' as the start of a conversation. It seems like the teacher made the most of every speaking opportunity for learners to practice. Meanwhile, visual aids were more often adopted in the intermediate classes, such as presenting pictures on the PowerPoint and asking learners to describe 'what do you see' in English; whereas students in the upper-intermediate course were more likely to be required to produce a short speech or oral presentation with PowerPoint demonstration through the 'share screen'. Moreover, students practised the speaking skills in two observed communicative activities. In the first activity, students asked one another about the provided questions listed under the pictures from PowerPoint for approximately 10 minutes. Following the activity, the feedback was observed, which entailed the students reporting their questions back to the teacher in the Zoom. The second observed communicative speaking practice was a role-play activity.

Both tasks involved leveraging the 'breakout room' feature in Zoom. This function was designed to split a large virtual classroom into smaller groups, and the host of the Zoom meeting room (in this case would be the teacher) would be able to decide how many rooms to assign out and how many participants per room. The teachers were allowed to hop between the different rooms to talk to the participants in the group. Meanwhile, there was an 'ask for help' button in every breakout rooms for calling an immediate group hopping from the teacher when students need an instant assistant. Curiously, it was noted that Alan (pseudonyms), the intermediate course teacher, seemed to avoid grammar correction during speaking practices and encouraged learners to articulate without interrupting them, while pronunciation corrections were captured in the upper-intermediate class. Finally, in all observed lessons, teachers' speaking speed was slower than normal, with different casual and light conversations in between the class. As such, it was observed that maintaining a good rapport with learners was also one of the priorities for teachers.

5.3 The Combination of TBLT, CLT and ICT

Figure 3.

TBLT + CLT		ICT	Frequency
- Assigned group work.	↔	Zoom 'breakout room'.	45
- Called on students to share ideas.	↔	'Share screen'; OneDrive.	34
- Checked on task progress during the group session.	↔	Group hopping; OneDrive	21
- Individual feedback and consulting time for assessments.	↔	'Consult hour' in private Zoom room.	8

As Figure 3 indicates, three teaching methodologies were able to coincide in a single lesson. In these online collaborations, teachers' most frequently used teaching technique echoed the group work with 'breakout room' feature in Zoom. Throughout, it was observed that the teacher broke the class into multiple individual 'breakout room' and each pair was given a specific task to complete after clarifying question instructions. Students worked in groups to communicate with peers and complete the task together. When the teacher was ready to bring students back to the main session for checking answers, the 'breakout rooms' would be closed with a one-minute timer in a pop-up window for students to rejoin the main session. This routine would be repeated by the teacher several times to accomplish the task thoroughly.

As well, the usage of OneDrive, another critical ICT tool, was also observed to have taken place during the whole group work interaction. Apart from the group hopping between different 'breakout room', teachers were able to co-author and collaborate on the documents which shared and used by learners through OneDrive. As such, if students or teacher make any changes or notes on the file, it would refresh with the updates made by the others. Each group has its own slide in the same PowerPoint file to type down the answers. Most of the time students divided the task into different parts and each member completed a part of the task individually in the 'breakout room'. It was rarely captured that students went through every question together.

In second observed upper-intermediate class, teacher introduced students with a new exercise, 'dictogloss', which often occurred in the live class (face-to-face) before. It was observed that the instructor began by clarifying the task rules, and subsequently, learners were split into different groups, following the task instruction. During the breakout room session, it was observed that teacher constantly checked on every group's progress through 'co-author' feature in OneDrive, and highlighted some answers in pink to signal the location of the errors. This process might repeat several times until students get it right. This allowed students to rethink and correct their mistakes with peers in the breakout room before being told the answers directly. This kind of instant feedback was less likely to be achieved in the traditional class, as normally teacher would not be able to pick up students' mistakes in details and allow learners to correct them before ending the physical group session. Also, teachers can monitor the 'live' progress when learners were typing, and even the names of multiple typewriters were indicated by OneDrive.

Afterwards, students were observed taking turns presenting answers as a whole group for approximately 10 minutes. Also, it was worth noting that every participant in the class shares the same vision, as an advance feature in OneDrive, and students could easily see the answers from different group's slide if clicking on another group's page. This might lead to invalid answers from weaker students, and the teacher participant claimed in the classroom that he was worried about this 'spy business' since two students were caught as their typewriter name appeared in other group's slide when they tried to copy the answers. At the end of the lesson, each student had equal opportunity to talk with the instructor individually in a private Zoom room, which aimed to go through student's works and answered questions from learners. This daily 'consult hour' session lasted approximately 1 hour in total, with 10 minutes for each student.

5.4 Note for Unusual Scene

Figure 4.

Types of Unusual Scene	Frequency
- Microphone;	6
- Internet;	7
- Virtual technology;	3
- ICT competence;	13
- Breakout room;	20
- No response from learners.	8

A few different types of unusual scenes were observed, as Figure 4 indicates. Throughout, microphone problems like experiencing noise, high pitched buzzing sounds, and strong echo occurred when participants spoke. Also, undesirable Internet conditions were evident as students disconnected and re-logged in multiple times during the lessons led to an inconvenient scene. Additionally, technologies might not always work under daily teaching routines. In a particular class, the teacher was unable to utilise the 'breakout room' function at that time and caused frustration for both the instructor and learners. It was observed that at the end the teacher had no choice but to abandon his original plan, which supposed to be group work, and the emphasis then shifted to practice comprehension reading and writing skills. This technology issue was finally solved by signing in again with IT team involvement. Again, a similar situation happened in the other class when all the student participants were ejected from the Zoom meeting as the teacher needed to restart the whole system. The reason was claimed afterwards by the teacher, as in: 'the bar with different features at the bottom disappeared, and the vision was stuck at one side when I try to split the screen between a word document and the Zoom. It jammed.'

Furthermore, the second most frequent issue Figure 4 has drawn attention to is the undesirable level of ICT literacy from our participants. It was captured 11 times that teachers forgot to unmute the microphone or were not connected to the audio feature while talking until students pointed this out. The teacher participant consequently complained that: 'there are too many features...sometimes I can't find the bottom.' Finally, the most frequent problem was often noted in the breakout room session. Students were told to work together as a group in the breakout rooms, while during the observations there was mostly silence or 'self-study mode' between group members. Also, the usage of learners' first language (not English) overwhelmingly dominated during the group sessions. It seemed quite challenging for the instructor to control the group work's quality as the nature of the 'breakout room' isolated students from the teacher's observation, whereas the instructor can see and hear from each group in a regular classroom.

It is possible that the type of task, the background of the learners, and the personality of the group members could lead to a different learning environment in the breakout room. For example, the 'silent mode' was seldom observed when the task was less academic-based and more on a daily-life basis. Also, students frequently use first language (L1) when they didn't even understand the question and what they were expected to do. In many cases, they might ask group members to translate for them. Moreover, personality could also be an important factor to create an effective group learning outcome, which students were more likely to stick with the 'English only policy' when the active learner steered the interaction and created cohesiveness within the group. In the end, there were a few times when the teacher called on students but did not hear back from them. Whether it was due to technical problems or other variables like negative learning altitudes it cannot be fully determined.

6. Discussion

Drawing on the data from the classroom observations, this study sought to discover what teaching methodologies and how they were employed in online courses to replace the traditional teaching pedagogies. In addressing these questions, we argue that the Task-based Language Teaching (TBLT) and the Communicative Language Teaching (CLT) were both adopted into the Information and Communication Technology (ICT) language teaching environment. TBLT was evident on the various online individual and group tasks and the feedback provided from instructors through OneDrive and Zoom, in line with the main characteristics of the TBLT mentioned by Horwitz *et al*. (1986) and Dim (2013). And CLT was evident on the oral presentation produced by learners and interactions and teamwork between students and teachers throughout the classes. Also, as discussed earlier, teacher in the intermediate class often followed this approach which encouraging students to make spontaneous speeches without concerning of grammatical accuracy. This echoes Savignon's finding (1991) that has indicated it helps improve learners' communicative competence, as the biggest advantage of CLT, when achieving fluency but accuracy in target language acquisition during the earlier stage. At the same time, Brooke (2011) reminds us that most group tasks have combined both TBLT and CLT, which equate with the Zoom breakout sessions in this study. For reducing bias and enforce the validity of the collected data, the use of those teaching methodologies was confirmed by the teacher participants. All of these lead us to conclude a 'relationship figure' between adopted pedagogy, as shown in the following table:

Figure 5. Relationship between ICT, TBLT and CLT in Online Environment

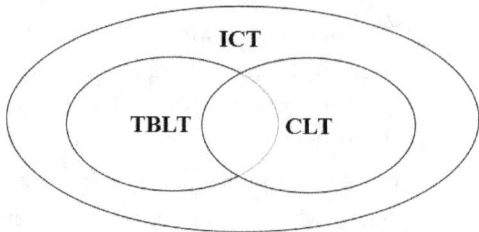

The elements of ICT were deployed served as a 'background' and played a 'media' role throughout the course, while TBLT and CLT were employed alternatively or used as an eclectic method for group tasks. Without ICT, none of these teaching methodologies was able to deliver under this COVID-19 crisis. Academics like Fotos and Browne (2013) admitted that the need for teachers to acquire technological skills and apply in teaching critically has never been greater. Language teachers must therefore embrace ICT as a powerful instructional tool and meet the challenge of this continually evolving academic world.

7. Limitations and Future Study

As for the limitations, it should be noted that the recruitment procedure of the participants in the present study is extremely challenging due to the COVID-19 restrictions. Due to the small number of the participants, the sample population is not truly representative and may have limitations as to how the results can be generalised. Also, the teacher participants expressed that being observed made them feel uneasy. The data collection process might have influenced teaching practices. For future study, it is recommended that the study be replicated with a larger number of participants which statistically represents the population, controlling for the potential confounding factors during the observation processes.

8. Acknowledgement

Many thanks go to the teachers and students for their time in participating in this project. I am grateful for the supports of the Language Centre at the University of Newcastle, Australia. Also, I would like to acknowledge the invaluable supports provided by my supervisor Dr. Alan Libert and the insightful comments from Dr. Christo Moskovsky.

9. References

Basilaia, G. and Kvavadze, D. (2020) Transition to online education in schools during a SA RS-CoV-2 coronavirus (COVID-19) pandemic in Georgia. *Pedagogical Research*, 5(4), pp.1-9.

Brooke, M. (2011) Teaching outside the comfort zone: Self-reflective practice in the ESL classroom. Paper presented at ACAL - Australian Council for Adult Literacy conference.

Creswell, J.W. (2002) *Educational research: Planning, conducting, and evaluating quantitative* (pp. 146-166). Upper Saddle River, NJ: Prentice Hall.

Dim, M. H. (2013) Developing the Task-based Strategies Syllabus to Enhance Communicative English Ability of Burmese Migrant Students. *Language in India*, 13(11).

Fotos, S., and Browne, C. M. (2013) *New perspectives on CALL for second language classrooms*. Routledge.

Horwitz, E. K., Horwitz, M. B., and Cope, J. (1986) Foreign language classroom anxiety. *The Modern language journal*, 70(2), 125-132.

Koondhar, M., MA, E., Siming, I. A., and Umrani, T. H. (2018) Language learning approaches: Unity in diversity. *Advances in Language and Literary Studies*, 9(6), 34-37.

Moorhouse, B. L. (2020) Adaptations to a face-to-face initial teacher education course 'forced' online due to the COVID-19 pandemic. *Journal of Education for Teaching,* pp. 1-3.

Owusu-Fordjour, C., Koomson, C. K., and Hanson, D. (2020) THE IMPACT OF COVID-19 ON LEARNING-THE PERSPECTIVE OF THE GHANAIAN STUDENT. *European Journal of Education Studies, 7(3), 88-100.*

Røkenes, F. M., and Krumsvik, R. J. (2016) Prepared to teach ESL with ICT? A study of digital competence in Norwegian teacher education. *Computers & Education*, 97, 1-20.

Sato, K., and Kleinsasser, R. (1999) Communicative Language Teaching (CLT): Practical Understandings. *The Modern Language Journal*, 83(4), 494-517.

Savignon, S. (1991) Communicative Language Teaching: State of the Art. *TESOL Quarterly*, 25(2), 261-277.

Shah, P. M., and Empungan, J. L. (2015) ESL teachers' attitudes towards using ICT in literature lessons. *International Journal of English Language Education*, 3(1), 201-218.

Thi Ly, T. T. (2018) Task-based language teaching: Teachers' perceptions and Implementation in the Australian ELICOS sector. *EA Journal*, 33(2).

Tomlinson, B., and Dat, B. (2004) The contributions of Vietnamese learners of English to ELT methodology. *Language teaching research*, 8(2), 199-222.

Tour, E. (2010) Technology use in ESL: An investigation of students' experiences and the implications for language education. *TESOL in Context*, 20(1), 5.

Yunus, M. M., Nordin, N., Salehi, H., Embi, M. A., and Salehi, Z. (2013) The Use of Information and Communication Technology (ICT) in Teaching ESL Writing Skills. *English Language Teaching*, 6(7), 1-8.

Distinct Linguistic Characteristics of Trakya Türkçesi: A Cultural Treasure

Devrim Yılmaz

University of New England, Armidale, Australia

ddevrim@une.edu.au

Distinct Characteristics of Trakya Türkçesi: A Cultural Treasure

Abstract: This paper presents findings from a linguistic analysis of a Turkish dialect spoken in Thrace (Trakya). The Trakya region, which is located at the very end of Southeast Europe, is shared by Bulgaria, Greece and Turkey; the part of the region in Bulgaria is known as Northern Trakya, the section in Greece is regarded as Western Trakya and the area in Turkey is considered Eastern Trakya. The Turkish dialects spoken in each section of the region share similarities. However, this paper focuses on the Turkish dialect spoken in Eastern Trakya. The dialect, Trakya Türkçesi or Trakyaca as it is known in Turkey, has distinct linguistic features. In order to explore this, I will refer to a segment of a YouTube video by a storyteller, who is from Eastern Trakya, present instances of distinct phonological, morphological and syntactic characteristics. Then, I will compare his language use against the standard Turkish variety (also known as Istanbul Turkish).

Keywords: Dialectology, Thracian Turkish, linguistics

1. Introduction

In this paper, I will present some of the distinct characteristics of Trakya Türkçesi or Trakyaca (Thracian Turkish) spoken in the region known as East Trakya. The Trakya region is shared by three countries; Bulgaria, Greece and Turkey; and the East Trakya region is the European part of Turkey contained by the Bulgarian-Turkish border, Greek-Turkish border and the natural coastal boarders of the Black Sea, the Sea of Marmara and the Aegean Sea. The following figure presents a map of Thrace.

Figure 1. Map of Trakya (East Thrace within Thrace, 2014)

Geographically, the East Trakya region consists of three provinces: Edirne, Tekirdağ, Kırklareli, and parts of two other provinces: Çanakkale (Gelibolu) and Istanbul. Linguistically, Thracian Turkish is spoken in provinces other than Istanbul due to the metropolitan status of Istanbul and the linguistic diversity of the Istanbul province. In this paper, I will discuss some of the distinct features of East Thracian Turkish in relation to phonology, morphology and syntax. In order to do this, I will have a closer look at some characteristics of Istanbul Turkish (Istanbul Türkçesi) and compare them against Trakyaca.

In the literature of Turkology, the standardised Turkish variety spoken in Turkey is referred as Turkey Turkish. Although referring to this variety as Turkey Turkish makes an important distinction between the Turkic languages spoken outside of Turkey; from a dialectological perspective, Turkey Turkish does not account for the rich variety within Turkey. Therefore, I will be using Istanbul Turkish to refer to the standardised variety of Turkish spoken in Turkey. The use of the term Istanbul Turkish is not free from issues either. Istanbul Türkçesi (Istanbul Turkish) historically refers to the variety spoken in Istanbul by the elite during the Ottoman Empire and Turkish Republic rule. This dialect served as the base for the language modernization reforms introduced by M. K. Atatürk during the early years of the Turkish Republic (1928 onwards). In other words, Istanbul Türkçesi has become the language of administration, education and mass media. In this paper, Istanbul Türkçesi will be used to refer to the standardised variety of Turkish spoken in Turkey rather than a regional dialect as Istanbul Turkish is used all over the country.

This paper explores some of the distinct characteristics of Trakyaca. The presented features of the dialect are presented with examples from Istanbul Turkish, and those examples will be followed by examples from Trakyaca. Linguistically speaking, both formal linguistics and systemic functional linguistics will be beneficial in the presentation and analysis of the examples. While formal linguistics, a.k.a. Chomskyan linguistics, mainly focuses on the form of linguistic units, constituents or categories, in other words, its focus is on the language itself (Chomsky, 1976); systemic functional linguistics, a.k.a Hallidayan linguistics, explores the ways in which linguistic units, constituents or categories function, in other words, it focuses on language use (Halliday, 1978). These linguistic schools will be used complimentarily and in a descriptive way in the presentation and analysis of the distinct features of Trakyaca with the aim of providing a framework for the richness of the dialect rather than defining linguistic phenomena prescriptively.

Some of the examples presented in this paper are taken from an interview where a storyteller, Kemal Dülger, who is telling childhood stories about his father and himself, focusing on his relationship with his father. The interview takes place at a TV show where Kemal tells his stories in Trakyaca and the host asks questions or make comments in Istanbul Türkçesi (Ara Kanal, 2019). The following sections will present examples and discuss the uniqueness of Trakyaca from phonological, morphological and syntactic perspectives.

2. Phonology

Phonology refers to the study of speech sounds of a language. Each language has distinct phonological properties and the dialects within each language have unique phonological properties as well. In this section, Trakyaca's uniqueness will be discussed in relation to vowel harmony, /h/ dropping and non-existence of /ğ/ (soft g).

2.1 Vowel Harmony

Vowel harmony is a distinctive characteristic of the Turkish language where vowels in words exhibit a particular pattern. Generally speaking, the vowels of any language can be categorised in relation to the place of articulation; whether they are articulated in the front section of the mouth or they are produced in the back section of the mouth close to the throat. Also, the vowels can also be studied in relation to their highness, in other words, whether they are produced close to the palate or the tongue. A broader and simplified definition of vowel harmony would state that back vowels /ı/, /u/, /o/, /a/ or front vowels /i/, /e/, /ü/, /ö/, /æ/ harmonise in Turkish. Particularly, a word should include only front vowels or only back vowels in order to be considered Turkish. If this is not the case, the word is either has its origins in other languages, e.g. Greek, Bulgarian, Armenian, Ladino, Farsi, Arabic or Hebrew, or it is Turkish but it went through phonological change in time. These words are regarded as exceptions. The following table presents a basic system for Turkish (Istanbul Türkçesi) vowels.

Table 1: Istanbul Turkish vowel system

	Front	Back
High	i	ı
Mid High	e, ü	u
Mid	ö	o
Low	æ	a

The Istanbul Turkish vowel system shows similarities with most languages, consisting of the basic five vowel sounds /i/, /e/, /u/, /o/, /a/. On top of that, Turkish has rounded front vowels /ü/, /ö/ and a high back vowel /ı/. This distinct phonological characteristic of the Turkish language sets out basic rules for vowel occurrence in root words (words without inflectional or derivational suffixes). It is also a way to determine whether a word has its origins in Turkish or in various other languages.

Another important feature of vowel harmony is that, due to the agglutinating nature of the language, which will be discussed in the morphology section, the suffixes that are attached to a root word must follow the rule. In other words, the vowel harmony can be observed in root words, as well as when suffixes are attached to a root word. The first set of examples are words (root) that follow vowel harmony. The first set of examples are from Istanbul Türkçesi and the second set of examples are from Trakyaca.

Example 2:

kedi (cat), *köpek* (dog), *kapı* (door), *kova* (bucket), *soğan* (onion)

As the examples show, the root words presented follow vowel harmony as each word consists of only front vowels or only back vowels. In the first two words, all the vowels are front vowels and the vowels in the other three words are back vowels. When these words are used in Trakyaca, they follow vowel harmony as well; however, the middle rounded vowels that appear word initial, /ö/, /o/, are replaced with their higher counterparts, /ü/, /u/. The examples below present the same words in Trakyaca.

Example 2:

küpek (köpek), *suuan* (soğan), *kuva* (kova)

As the examples show, the root words (kök) presented follow vowel harmony. However, in the following set of examples, the quality of the initial front vowels changes. In Istanbul Türkçesi, they are lower, but they become higher in Trakyaca while the vowel harmony is preserved despite the shift in vowel quality. This phonological process is unique to Trakyaca and might have its origins in the rich linguistic repertoire of the Trakya region, which has been home to various peoples and languages.

Another phonological characteristic of Trakya Türkçesi is also regarding vowel harmony, however, it is slightly different from the vowel harmony observed in root words or when suffixes are attached. A unique phonological feature in relation to vowel harmony is present at clause level in Trakyaca. The vowel harmony in Istanbul Türkçesi is about word level including the root word and suffixes. However, Trakyaca has examples of vowel harmony at clause level. The following example sheds light on this feature.

Example 3:

Istanbul Türkçesi: *Ne yap-ıyor-sun*? (What are you doing?)

Trakyaca: *Naabuyun*?

The clause "what are you doing" has two words: *ne* and *yapıyorsun*. In Istanbul Türkçesi, vowel harmony only applies to words. *Ne* is a monosyllabic word and has only one vowel. *Yapıyorsun* can be broken down into its constituents:

yapıyorsun =

yap (root: meaning to do, to make) +

-ıyor (present progressive inflection) +

-sun (second person singular inflection).

The same clause becomes *naabuyun* in Trakyaca. Firstly, two words, *ne* and *yapıyorsun* merge into one and they become *naabuyun*. Secondly, the front vowel /e/ in the word *ne* transforms to the back vowel /a/, and the merged word follows vowel harmony consisting of only back vowels.

2.2 /h/ dropping

The dropping of /h/ is another phonological characteristic of Trakyaca. The syllable initial or syllable final /h/ is dropped in Trakyaca similar to some varieties of English. The dropping of /h/ is a universal phonological process, although it is not the characteristic of Istanbul Türkçesi. The following set of examples present words with the sound /h/ in Istanbul Türkçesi and the dropped /h/ in Trakya Türkçesi.

Example 4:

Istanbul Türkçesi: havlu, hava, saha, semah

Trakyaca: avlu, ava, saa, semaa

The words *havlu* (towel), *hava* (air, weather) and *saha* (field) have word/syllable initial /h/ sound and *semah* (Alevi Bektashi ritual) has syllable final /h/ sound and they are all pronounced in Istanbul Türkçesi. However, the /h/ is deleted in Trakyaca, which makes it a unique characteristic of the dialect. Similar to the distinct vowel harmony related phonological characteristics of Trakyaca, the deletion of word/syllable initial /h/ might have its roots in the linguistic repertoire of the region including the Pomak language, Greek, Romanca, Macedonian, Albanian and Ladino among others.

2.3 Non-existence of /ğ/

In Istanbul Türkçesi, /ğ/ (soft g) can be considered a glide like sound. Glides are considered semi-vowels as they sound like vowels but function as consonants. For example, the sound /y/ is considered a glide, which has both vowel-like and consonant-like characteristics. /y/ sounds rather similar to the vowel sound /i/ but can function as a "blending sound" (kaynaştırma sesi). Another characteristic of /ğ/ is that it never occurs word/syllable initial. The following set of examples present words with /ğ/.

Example 5:

Istanbul Türkçesi: *dağ* (mountain), *sağ* (alive, right), *çağ* (era, period)

Trakyaca: daa, saa, çaa

These set of examples present words ending with "soft g". All these words follow the same pattern of CVC (consonant, vowel, consonant). In Istanbul Türkçesi, "ğ" is pronounced and the pronunciation is a distortion in the quality of the preceding vowel.

However, the vowel which precedes the "ğ" is lengthened in Trakyaca. This pattern is also observed when the words ending with "ğ" receives a suffix (ek). The following set of examples provide simple clauses with inflected words with /ğ/ in the final position.

Example 6:

Istanbul Türkçesi: *Biz Tekirdağ'a gittik.* (We went to Tekirdağ.) *Araba sağa döndü.* (The car turned right.)

Trakyaca: Tekirdaaya gittik. Araba saaya döndü.

In both of these clauses, the words *dağ* and *sağ* have an inflectional suffix "-e/a" (dative case – ismin -e hal eki), which follows the root word. However, due to the non-existence of /ğ/ in Trakyaca, the preceding vowel is lengthened (pronounced as double /a/). As the word ends with a vowel in order to attach the dative case suffix "-e/a" requires a blending sound, which is a /y/ in this case (see beginning of this section). Both examples of *daaya* and *saaya* provide linguistic evidence for the non-existence of "ğ" as a distinct sound in Trakya Türkçesi. As these words have distinct qualities and end with a vowel, they require a blending sound (*daa-y-a, saa-y-a*).

3. Morphology

Agglutination is an important characteristic of Turkish. The etymology of the word goes back to its Latin origins and it means "to glue" (Merriam-Webster, 2020). From a linguistic perspective, the process is "to glue" inflectional or derivational morphemes to word roots. The following word *Edirne* is "glued" by various morphemes.

Example 7:

Edirne-li-leş-tir-ebil-dik-lerimiz-den-dir-ler = 1 word

(They are the ones that we could turn into Edirne residents.) = 11 words in English

This word can be broken down to its constituents in the following way. Each constituent is presented with morphological information.

Edirne = root word – name of province in East Trakya bordering Greece and Bulgaria

-li = derivational suffix that forms adjectives from nouns

-leş = derivational suffix that forms verbs from nouns

-tir = derivational suffix that forms causative verbs

-ebil = inflectional suffix for modal verbs

-lerimiz = inflectional suffix for third person plural possessive

- den = inflectional suffix for dative

- dir = inflectional suffix for present tense

- ler = inflectional suffix for third person plural

As shown in the analysis, each suffix has a unique form and function. While the form can change due to vowel harmony, the function of these morphological constituents remains the same. The following set of examples, on the other hand, present one of the unique morphological characteristics of Trakyaca.

Example 8:

Istanbul Türkçesi: *merhaba* (hello), *haydi* (come on), *güle güle* (bye bye)

Trakyaca: *merabayın* (hello to you all), *aydiyin* (come on you all), *güle güleyin* (bye bye to you all)

In Istanbul Türkçesi, the same words are pronounced as *merhaba*, *haydi* or *güle güle* and they don't require any agglutination as they are categorised as "greeting and goodbye" words. However, the same words can be used with the suffix for the second person plural imperative with a totally different function. The following examples show how the second person plural imperative is used.

Example 9:

Yap-ın = *yap* (root verb meaning do, make) + *-ın* (second person plural inflection)

Ara-y-ın = *ara* (root verb meaning look for, search) + *-y* (blending sound) + *-ın* (second person plural inflection)

The suffix *-ın/in* (depending on vowel harmony) is attached to root verbs in order to indicate the person who is supposed to carry out the order. While one person receives an order only the root verb, e.g. *yap* (do, make), *ara* (search), is used. When more than one person is instructed, a personal inflection is attached to the verb. In this case, it is *-ın/in*. The second person plural imperative inflection is only used in this context in Istanbul Türkçesi. However, the suffix is also used in a completely different way in Trakyaca in order to construe plurality with addressing words (see Example 8).

4. Syntax

The last section in this paper is the flexible word order patterns of Trakyaca. Syntactically, Istanbul Türkçesi is known to follow a S-O-V (subject-object-verb) structure, and the sentences in this word order are considered rule-governed. However, Trakyaca does not follow this order. It is quite common to hear a sentence starting with the verb and then continuing with the object. The subject might come at the end or does not take place at all. The following sentences are from the Ara Kanal YouTube video.

Example 10:

Trakya Türkçesi: Ama şindi lastik patlamış bunların yolda gelirken. (Ara Kanal, 2019)

Istanbul Türkçesi: *Ama babamlar yolda gelirlerken, lastikleri patlamış.* (While they were driving back, they had a flat tire.)

Trakya Türkçesi: Beş metre iple baaladılar pederi sırtıma. (Ara Kanal, 2019)

Istanbul Türkçesi: *Babamı sırtıma beş metre ip ile bağladılar.* (They tied my father to my back with a five metre long rope.)

The first sentence, which is taken from the video, is representative of the word order that does not follow the S-O-V rule. The verb, *patlamış* (burst – past participle) is placed in the middle of the clause and the noun group, *bunların lastik*, is separated in the middle by the verb, *patlamış*. In Istanbul Türkçesi, the same noun group/phrase would be kept together as *bunların lastik* (their tyre) and the verb would come at the end of the clause. Similarly, the verb *baaladılar* (they tied) is placed in the middle of the clause in Trakyaca. However, the verb is expected to be placed at the end of the sentence in Istanbul Türkçesi. This unique characteristic of Thracian Turkish creates a warm and friendly discourse in comparison to the rule governed standardised variety: Istanbul Turkish.

5. Conclusion

This paper presented some of the distinct linguistic characteristics of Trakyaca in relation to phonology, morphology and syntax. Although each of these linguistic layers, even distinct characteristics within each linguistic layer, can be analysed and presented separately; I aimed to provide a general picture, for the readers who are not familiar with Turkish or linguistics in general, in relation to the way Trakyaca sounds and used in relation to morphological and syntactical processes. The characteristics discussed in relation to phonology were the unique use of vowel harmony, /h/ dropping and non-existing /ğ/ (soft g). In terms of morphology, a unique morpheme use was discussed. Finally, the flexible word order of Trakyaca was presented and discussed in comparison to Istanbul Türkçesi. According to Holmes & Wilson (2013), the standardised variety of a language is generally considered to be the high variety and the rest are low varieties, but this distinction remains to be a point of debate. In our context, although Istanbul Türkçesi might be considered the high variety of Turkish by some; the "low varieties" embody so much cultural, social and linguistic exuberance. Therefore, Thracian Turkish needs to be acknowledged, maintained and celebrated as one of the important cultural treasures of South Eastern Europe.

References

Ara Kanal (2019, October 10). *Yediği Dayaklar Full – Babanın Oglu Kemal.* [Video file]. YouTube. Available at: https://www.youtube.com/watch?v=96BifrREMu0&t=274s&ab_channel=ARAKANAL [Accessed 19 Nov. 2020].

Holmes, J. and Wilson, N. (2013). *An Introduction to Sociolinguistics.* New York: Routledge.

Chomsky, N. (1976). *Reflections on Language.* London: Temple Smith.

Halliday, M. A. K. (1978). *Language as a Social semiotic: The Social Interpretation of Language and Meaning.* London: Arnold.

East Thrace (blue) within Thrace [Online image]. (2014). Wikipedia. https://en.wikipedia.org/wiki/East_Thrace#/media/File:EstThraceWithinThrace.png [Accessed 19 Nov. 2020].

Merriam-Webster. (2020). Agglutination. In Meriam-Webster.com dictionary. Available at: https://www.merriam-webster.com/dictionary/agglutinate#etymology [Accessed 19 Nov. 2020].

Establishing Rapport with Evaluative Language in Online Hotel Responses

Ly Wen Taw

School of Humanities and Social Science, Faculty of Education and Arts, The University of Newcastle, Australia

Department of English, Faculty of Modern Languages and Communication, Universiti Putra Malaysia, Malaysia

Centre for the Advancement of Language Competence (CALC), Universiti Putra Malaysia, Malaysia

Taw.LyWen@uon.edu.au
lywen@upm.edu.my

Establishing Rapport with Evaluative Language in Online Hotel Responses

Abstract: Tourism, particularly cultural heritage tourism, promotes the conservation and preservation of a country's cultural and natural heritage. The hospitality industry plays a vital part in the development of tourism. In this digital era, electronic Word-of-Mouth (eWOM) has gained popularity with the development of the internet and has significantly influenced consumers' purchase decisions. In the hotel industry, there has been an increase in the use of social media that has led to the emergence of various websites with online hotel reviews, such as *TripAdvisor*. Given the considerable influence of eWOM, hotel responses have become increasingly essential to positively influence consumers' purchase decisions. Online follow-up customer service, such as responding to customers' online reviews, is an effective way of reaching customers and engaging in online reputation management. The positive emotions from the management have been widely known to enhance rapport with customers. Building on Appraisal Theory (Martin & White, 2005), this study explores the use of evaluative language by the hotel management in Malaysia in responding to negative online reviews on *TripAdvisor* to establish rapport with their customers. The evaluative language focuses on the linguistic resources in the *affect* sub-system of the theory, which are utilised for expressing positive and negative emotions. The data was collected from 5-Star, 4-Star, and 3-star hotels in three different destinations. The study findings show hotels in these three categories had a strong preference towards positive *affect* evaluations. Among the three hotel star categories investigated, the 3-Star hotels recorded the lowest frequencies in both positive and negative *affect* evaluations. Interestingly, the 4-Star hotels had the highest occurrences of negative *affect* evaluations. The 5-star hotels recorded the highest occurrences of positive *affect* evaluation. An examination of hotel responses to negative reviews will establish an understanding of evaluative language used by the hotel management in Malaysia to establish rapport with the customers.

Keywords: evaluative language, online reviews, hotel responses, rapport

1. Introduction

Tourism is not only vital in the growth of the economy at both national and global scales, but it also contributes significantly to the protection and preservation of natural, historical, and cultural heritage. Tourism provides the positive and lasting effects on our cultural and natural heritage assets (Robinson & Picard, 2006). Tourism contributes significantly to Malaysia's economy, as the tourism sector is the third highest source of foreign income after manufacturing and palm oil industries. Malaysia is recorded as the third most visited country in Asia after China and Thailand, with 26.75 million international tourists arriving in 2016 (World Tourism Organization, 2018).

The tourism industry correlates closely with the hotel industry, as Johnson and Vanetti (2008) emphasise, and it is an essential sub-sector of the tourism industry. The Department of Statistics Malaysia (2019) demonstrates that the accommodation contributes considerably to tourism expenditures, after shopping and transport services. Padlee, Thaw, and Zulkiffli (2019) point out that the hotel industry has become one of the important sources of revenue in Malaysia's tourism industry.

As the internet has dramatically revolutionised many aspects of life in this digital age, there has been an increase in the prevalence of electronic Word-of-Mouth (eWOM) in the hotel sector. EWOM is the communication between consumers on online platforms about a product or service provided by companies. The influence of eWOM seems potent to both consumers and companies. Litvin, Goldsmith, and Pan (2008) contend that the influence of eWOM has become increasingly crucial in the hospitality and tourism industries. As the effects of online reviews and eWOM can be profound, hotel management's responses to the customers' online reviews can influence positively on the reputation of the hotel and customers' purchase decisions.

Positive affectual expressions from service organisations are known as an effective way to enhance customer relationships (Wang et al., 2017). This circumstance can be termed as emotional contagion. Emotional contagion in customer relationship influences significantly on customer satisfaction during the interaction with customers (Barger & Grandey, 2006). With the increasing use of the internet in this global age, many individuals are exposed to various emotion expressions in the digital realm, thereby resulting the occurrence of emotional contagion, which is known as digital emotional contagion. This digital emotional contagion appears to have a powerful effect on internet users' emotions (Goldenberg & Gross, 2020), so emotional contagion in the employee-customer interactions can also be mediated by electronic means.

With the notion of emotional contagion that can enhance the rapport with customers, this study examines the evaluative language of positive and negative emotions used by the hotel management of 5-Star, 4-Star, and 3-Star hotels in Malaysia in responding to customers' negative reviews to establish rapport with customers.

2. Theoretical framework: Rapport Management Model (RMM)

Language has its vital role in developing rapport in social relations. With the Rapport Management Model (RMM), this study focuses on the use of evaluative language that can establish connections. Spencer-Oatey(2008) proposes the theoretical framework to examine the ways of using language to build and maintain rapport in social interactions.

There are three bases in the model: face sensitivities, sociality rights and obligations, and interactional goals. The first element of the framework is face sensitivities. Spencer-Oatey

(2000) states that there is a close relation between face and an individual's sense of identity or self-concept. The second basis of rapport is sociality rights and obligations, in which people perceive themselves to have a range of sociality rights and obligations when relating to others. Finally, an interactional goal is the third element in the model that can affect interpersonal rapport. This element is related to people's specific goals in interactions with others, and they can be relational, transactional in nature or task-focused.

Spencer-Oatey (2000) maintains that rapport orientation is the one of the major factors in rapport management strategies. There are the four types of rapport orientations, which are: rapport enhancement orientation, rapport maintenance orientation, rapport neglect orientation, and rapport challenge orientation. The former two can strengthen rapport, while the latter two can jeopardise rapport.

Spencer-Oatey (2008) identifies the rapport management strategies from a linguistic perspective in these five domains: illocutionary, discourse, participation, stylistic, and non-verbal domains. This study focuses on one of the domains - stylistic domain to examine the lexical choices in expressing emotions by the hotel management when responding to customers' negative reviews. The next sub-section will explain further the theory that shapes the data analysis in the stylistic domain.

2.1 Stylistic domain: Appraisal Theory

Within the stylistic domain, Spencer-Oatey (2008) asserts that the choice of lexis can have considerable impact on interpersonal relations. To analyse the stylistic domain of the hotel responses, Appraisal Theory (Martin and White, 2005) was applied to examine the evaluative language used to build and maintain the rapport with the customers through online reviews on the TripAdvisor online community.

Appraisal Theory is a set of a system of evaluative resources in language. Appraisal is defined as a "linguistic resource used to construct interpersonal meaning" (Martin & White, 2005, p.35). In Appraisal Theory, evaluative resources are divided into three basic systems of semantics: Attitude, Engagement, and Graduation. The focus of this study is the sub-system of Attitude, and the next sub-section will present one of its sub-systems, affect.

2.2 Appraisal Theory: *Attitude-affect*

Attitude is the central system of Appraisal Theory, which leads the data analysis of the stylistic domain of this study. This system comprises the three semantic regions embodying emotion, ethics, and aesthetics. In other words, *Attitude* entails the expressions of human emotions, as well as the evaluation of behaviour, personalities, objects, and events. According to Martin and White (2005), the system of *Attitude* is classified into three sub-systems, which are *affect*, *appreciation*, and *judgment*.

As the research question of the study focuses on expressions of emotions by hotel management in rapport management with customers, the sub-system of *affect* was singled out to examine the polarity of positive and negative feelings expressed in the responses. Martin and White (2005) maintain that feelings are construed as the realisations of qualities, mental and behavioural processes, and modal adjuncts as illustrated as below:

- *Affect* as qualities: the *happy* customer
- *Affect* as mental process: the customer *loves* the service provided.
- *Affect* as behavioral process: the customer *compliments* the staff and manager.
- *Affect* as modal adjuncts: *Happily*, the customer gave the positive online review.

[Adapted from Martin & White (2005)]

Martin and White (2005) further categorise the sub-system of affect into the semantic topology of affect emotions groups between positive and negative polarities: satisfaction, security, happiness, and inclination, and all these groups are presented with the examples in the following table.

Table 1: *Affect* sub-system of Appraisal Theory

Kinds of *Affect* Sub-system	Semantic Typologies	Positive (+)	Negative (-)
Satisfaction (+) / dissatisfaction(-)	(+) interest, pleasure (-) ennui, displeasure	pleased, impressed, reward	angry, bored, scold
Security (+) / insecurity(-)	(+) confidence, trust (-) worry, surprise	confident, assured, entrust	uneasy, anxious, freak out
Happiness(+) / unhappiness(-)	(+) cheer, affection (-) misery, antipathy	cheerful, love, adore	gloomy, dejected, weep
Inclination(+) / disinclination(-)	(+) desire (-) fear	miss, long for, yearn	wary, fearful, tremble

3. Methodology

The data were collected over three months from January 2020 to March 2020, from a travel online reviews website—*TripAdvisor*. *TripAdvisor* is known as the world's largest travel site, with more than 830 million online reviews (Kinstler, 2018; TripAdvisor, 2017). *TripAdvisor* allows two-way communication between customers and management representatives, who represent accommodation venues, restaurants, or attractions; and the latter can respond to the posted online reviews by the customers or travelers. The online reviews are categorised into five traveler ratings from *Excellent, Good, Average, Poor*, to *Terrible*.

The hotel responses from 5-star, 4-star and 3-star hotel rating categories in the selected destinations in Malaysia were collected for analysis. To ensure the finding validity, this study applies the data triangulation technique to its sources. Thus, six hotels in each of three destinations: Kuala Lumpur (KL), Selangor, and Pahang were selected from among the thirteen states and three federal territories in Malaysia. A total of eighteen hotels were chosen. The hotel industry correlates significantly with tourism destinations (Attila, 2016), so the selected hotels in these three destinations are all popular tourist destinations in Malaysia.

Purposeful sampling, which is defined as selection of "information-rich cases" to illustrate the central importance in the purpose of the study (Patton, 1990, p. 169), was adopted in this research. It is considered as the sampling design that can maximise the range of variation (Palinkas et al., 2015). Applying the purposeful sampling in this study, the responses to the negative reviews, which consist of traveler ratings of either *Poor* and *Terrible,* were selected from 5-Star, 4-Star, and 3-Star hotels in KL, Selangor, and Pahang. Two responses for the negative reviews of *Poor* and *Terrible* traveler ratings were selected for each of the eighteen hotels in this study. In other words, there were four responses collected from each hotel, giving a total of 72 responses to the negative reviews in the data collection. Due to the selected sampling method, some responses were written in 2018, although the data collection occurred from January to March 2020.

Using NVivo 12, the affectual instances in positive and negative polarity were coded according to the hotel star rating in the responses. The affectual evaluations were quantified in term of the frequency use, and the evaluations were analysed qualitatively with reference to the bases of the rapport in the theoretical framework, along with its rapport orientation.

4. Results and discussion

Figure 1 demonstrates the frequency of affectual use in polarity among the hotels of different rating in Malaysia. As shown in the chart, positive *affect* instances considerably outnumbered the negative ones. 5-Star hotels (N=163) were recorded as the highest frequency of positive affectual use, and it was followed by 4-Star hotels and 3-Star hotels. However, for the negative affectual evaluation, the 4-Star hotel category had the highest frequency (N=29). The next sections will present the positive and negative polarity in affectual evaluations.

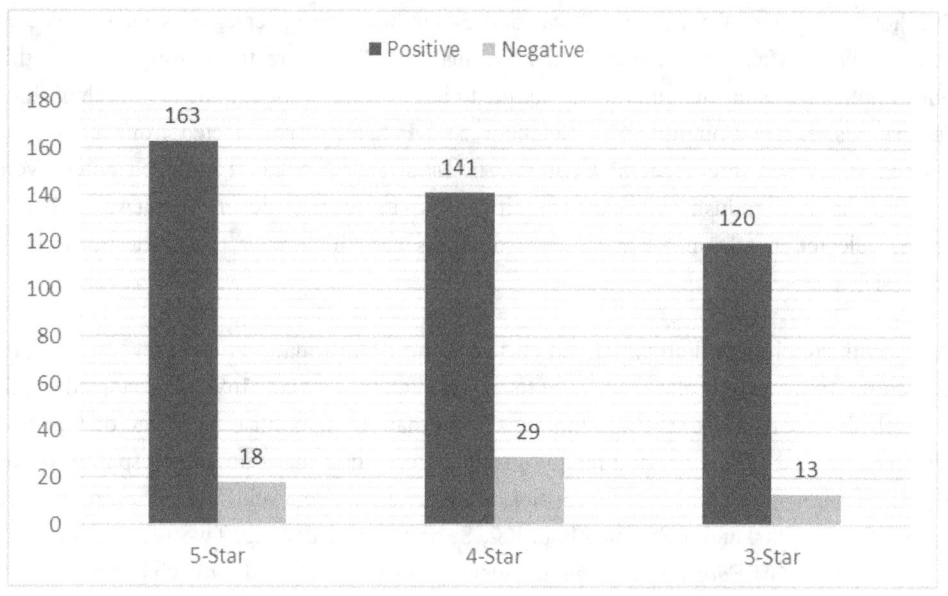

Figure 1: Frequency of *affect* instances in polarity among the hotel star rating categories

4.1 Use of positive affect in responding negative reviews

To establish rapport with customers, the hotel management predominantly engaged in positive affectual evaluations in responding customers' negative reviews. Table 2 illustrates the positive *affect* types employed by the different hotel star rating categories.

Table 2: Positive affect *types of hotel star rating categories*

Affect Types (positive)	5-Star	4-Star	3-Star	Total
Happiness	21	9	10	49
Inclination	45	55	51	151
Satisfaction	51	56	41	142
Security	46	22	18	85
Total	163	141	120	427

A Chi-Square test of independence was performed to examine the statistical difference between the variables. It was found that there was a significant relationship (p=0.001, p<0.05) between the hotel star rating and the positive affectual evaluation. Here are some samples of the *affect* types used by the hotels of different star rating:

Happiness

- We will do our best to give you the great hotel experience that so many of our guests have <u>grown so fond of</u>. (*5-Star*)
- We hope to afford the opportunity to <u>welcome you back</u> on your next travelling journey. (*4-Star*)

Inclination

- We are very <u>sorry</u> and <u>apologize</u> for the inconvenience cause by this incident happen. (*5-Star*)
- We <u>look forward</u> to your return with your group. (*3-Star*)

Satisfaction

- We are very <u>thankful</u> for your comment. (*4-Star*)
- We do <u>appreciate</u> your feedback. (*3-Star*)

Security

- We would like to <u>guarantee</u> you that we will look into the importunities to improve. (*5-Star*)
- Rest <u>assured</u> necessary steps has been taken to rectify the issue. (*3-Star*)

In the topology of the positive *affect* types, inclination was the manifestation of positive evaluations by the hotels. Martin and White (2005) state that inclination refers to the emotion expression when the intention is involved, rather than reaction. As seen in the examples above, the affect mental state of "sorry" and behavioural surge of "apologize" were categorised in the positive polarity of affectual evaluation in inclination.

It is crucial to place these words in the context of the study in which hotel management responded to the customers' negative reviews. The emotive state of being sorry and behavioural process of apologising imply the desire to establish positive relations. It is essential to pay attention to the contextual evaluative roles of the aforementioned affective mental state and behavioural process that can lead to a positive connection. Therefore, being "sorry" and the act of "apology" appear to entail admission of fault, so they are considered as rapport enhancement orientation, as it can rebuild the trust with customers in the study context by attending to the dissatisfied customers' face wants.

The semantic use in the affective types of happiness is oriented in line with rapport enhancement with the customers. As shown in the example—"welcome you back", the *affect* type of happiness was realised in the surge of behaviour in affection towards customers by making them feel appreciated and welcome. This affective instance occurred the most frequently among the semantic use of affectual happiness. The hotel management intended to establish the relation with the customers by creating the sense of appreciation and gratitude in them with the *affect* type of satisfaction.

To repair the relationship with dissatisfied customers, the hotel management used the affectual language such as "guarantee" and "ensure" which are realised in the behavioural surge in security to earn customers' trust and retain loyal customers. These aforementioned positive *affect* types attend to customers' wants and sensitivities that indirectly emphasise customers' value and worth to the hotels. The next sub-section explores the use of negative *affect* in rapport management in the responses made to customers' negative reviews.

4.2 Use of negative affect in responding to negative reviews

Although the hotel management had a preference towards positive affectual evaluations, hotels of different star ratings also used negative affectual evaluations in the responses to the negative reviews. Table 3 illustrates the distribution of different negative *affect* types for 5-Star, 4-Star, and 3-Star hotels.

Table 3: Negative affect *types of hotel star rating categories*

Affect Types (negative)	5-Star	4-Star	3-Star	Total
Unhappiness	6	5	1	12
Disinclination	0	0	0	0
Dissatisfaction	9	22	12	43
Insecurity	3	2	0	5
Total	18	30	14	61

As can be seen in Table 2, the data for the negative *affect* appears potentially non-normal distributed, as the values for some of the negative *affect* types were less than 5, or even 0. Therefore, Chi-Square one-variable test was used to investigate the statistical difference between the variables. The result indicates that there is a statistically significant difference ($p=0.023$, $p<0.05$). In other words, the use of negative affectual evaluation was significantly different between 5-Star, 4-Star, and 3-Star hotels.

Table 3 illustrates that the negative pole of *dissatisfaction* was recorded as the highest frequency among other negative *affect* types. The negative *affect* types used by the hotels are exemplified as follows:

Unhappiness

- We are <u>disheartened</u> to read that your stay with us was not up to your expectations. (5-Star)
- It is always <u>sad</u> for any manager to read when guests have not enjoyed their stay. (4-Star)

Dissatisfaction

- I very much <u>regret</u> the inconvenience you experienced. (4-Star)
- I would like to personally extend my sincerest apologies for the <u>disappointment</u>. (3-Star)

Insecurity

- We are disappointed and <u>disturbed</u> to learn of your unpleasant experience during your recent stay with us. (5-Star)
- I am sure that there was never any intention of <u>yelling</u> or make you feel <u>scared</u>. (4-Star)

The lexical choice in the negative *affect* types in *unhappiness*, *dissatisfaction*, and *insecurity* are realised as the negative *affect* types because of the definitions that they carry which do not seem to achieve the positive rapport with customers. Thus, placing the lexical choice in the study context, the lexical use in these semantic regions appear to incline towards rapport-neglect orientation.

Expressing emotions of *unhappiness*, *dissatisfaction*, or *insecurity* might be intended to convey empathy. However, it only appears to neglect the customers' dissatisfaction, as dissatisfied customers usually expect an apology and promise of improvement, or offer of repair. In other words, the negative affectual language did not seem to attend to the customers' social entitlement. For example, when the hotel management responded to negative reviews by expressing unhappiness as shown in the example: *"it is always sad for any manager to read when guests have not enjoyed their stay"*, it did not acknowledge the customers' expectations of the improved service, despite the attempt to allay the customers' concern.

On the other hand, the expression of regret is considered to be different from "sorry" because expression of regret simply implies the recognition of the service failure. Apart from that, it conveys a dispassionate tone and seems to preclude the acceptance of responsibility. In general, dissatisfied customers regard themselves as having sociality rights as the hotel clients and develop the behavioural expectation of their perceived sociality right. In other words, it can be concluded that the negative affectual language use did not fulfil sociality rights and obligations of the rapport management, which can affect the interpersonal relation between the customers and management.

5. Conclusion

Since this theory is the major theoretical foundation of the study, the findings can expand the existing literature on rapport management in workplace communication, particularly in hotel management, tourism, and hospitality industry. From the findings, it was found that some hotel management could have communicated with the dissatisfied customers with more positive evaluative language for the service recovery. Business-related ESP courses should raise awareness of the evaluative language that focuses on the positive words instead of the negative ones. Given the importance of digital emotional contagion via eWOM, positive language related to expressing feelings should be used when interacting with customers.

References:

Attila, A. T. (2016). The Impact of the Hotel Industry on the Competitiveness of Tourism Destinations in Hungary. *Journal of Competitiveness, 8*(4), 85-104. doi:10.7441/joc.2016.04.06

Barger, P. B., & Grandey, A. A. (2006). Service With a Smile' and Encounter Satisfaction: Emotional Contagion and Appraisal Mechanisms. *Academy of Management Journal, 49*(6), 1229-1238.

Department of Statistics Malaysia. (2019). TOURISM SATELLITE ACCOUNT 2018 [Press release]. Retrieved from https://www.dosm.gov.my/v1/index.php?r=column/pdfPrev&id=Wk1KWlpxZTRDWnVhVWNMV21ZVVY3Zz09

Goldenberg, A., & Gross, J. J. (2020). Digital Emotion Contagion. *Trends in Cognitive Sciences, 24*(4), 316-328. doi:https://doi.org/10.1016/j.tics.2020.01.009

Johnson, C., & Vanetti, M. (2008). Internationalization and the Hotel Industry. In A. G. Woodside & D. Martin (Eds.), *Tourism Management: Analysis, Behaviour and Strategy* (pp. 285-301). Oxford, UK: CAB International

Kinstler, L. (2018). How TripAdvisor changed travel. Retrieved from https://www.theguardian.com/news/2018/aug/17/how-tripadvisor-changed-travel

Litvin, S. W., Goldsmith, R. E., & Pan, B. (2008). Electronic word-of-mouth in hospitality and tourism management. *Tourism Management, 29*(3), 458-468. doi:10.1016/j.tourman.2007.05.011

Martin, J. R., & White, P. R. R. (2005). *The Language of Evaluation: Appraisal in English.* New York: Palgrave Macmillan.

Padlee, S. F., Thaw, C. Y., & Zulkiffli, S. N. A. (2019). The relationship between service quality, customer satisfaction and behavioural intentions. *Tourism and hospitality management, 25*(1), 121-139. doi:10.20867/thm.25.1.9

Palinkas, L. A., Horwitz, S. M., Green, C. A., Wisdom, J. P., Duan, N., & Hoagwood, K. (2015). Purposeful Sampling for Qualitative Data Collection and Analysis in Mixed Method Implementation Research. *Administration and policy in mental health, 42*(5), 533-544. doi:10.1007/s10488-013-0528-y

Patton, M. Q. (1990). *Qualitative evaluation and research methods.* Beverly Hills, CA: SAGE Publications.

Robinson, M., & Picard, D. (2006). *Tourism, Culture and Sustainable Development.*

Spencer-Oatey, H. (2000). *Culturally Speaking: Managing Rapport Through Talk Across Cultures*: Continuum.

Spencer-Oatey, H. (2008). *Culturally speaking: culture, communication and politeness theory*: Continuum.

TripAdvisor. (2017). About TripAdvisor. *Media Center.* Retrieved from https://tripadvisor.mediaroom.com/US-about-us

Wang, Z., Singh, S. N., Li, Y. J., Mishra, S., Ambrose, M., & Biernat, M. (2017). Effects Of Employees' Positive Affective Displays on Customer Loyalty Intentions: An Emotions-As-Social-Information Perspective. *Academy of Management Journal, 60*(1), 109-129.

World Tourism Organization. (2018). *UNWTO Tourism Highlights 2018 Edition.* Retrieved from https://www.e-unwto.org/doi/pdf/10.18111/9789284419876

Distinguishing Between Language Difference and Language Disorder in Deaf Children who use Signed Language

Joanna Hoskin[1], Hilary Dumbrill[2], Wolfgang Mann[3]

[1] *City, University of London, UK*

[2] *Hamilton Lodge School and College, Brighton, UK*

[3] *University of Roehampton, UK*

joanna.hoskin@city.ac.uk

hilary.dumbrill@hamiltonlsc.co.uk

Wolfgang.Mann@roehampton.ac.uk

Distinguishing Between Language Difference and Language Disorder in Deaf Children who use Signed Language

Abstract: This paper provides an overview of the use of dynamic assessment (DA) with deaf children who use signed language. It starts with background information about deaf children, including their language learning experiences, the Deaf community, and the community language culture. This provides the context for why the use of DA is important for this group of children. In order to link these background topics for a clinical or educational setting, a practitioner's dilemma when assessing a deaf child's language is described next. A brief overview of dynamic assessment and its history is then given. Two case studies provide examples which link theories with practical application of the DA processes. The paper concludes with a summary of the next steps needed to share DA techniques and ensure deaf children's language is assessed within the context of their individual language learning environments. This includes taking account of the skills of the child's communication partners, their competence in using mediated learning techniques with the child, and the child's individual ability to benefit from intervention and demonstrate changes in their language use. This collaborative work on DA between researchers and practitioners provides a model for the iterative process of everyday practice, informing the evidence seeking process via research.

Keywords: deaf children; dynamic assessment; signed language; researcher/practitioner collaboration

1. Introduction

The authors of this paper have been working together for several years to develop evidence-based practice, practice-based evidence, accessible information and child-focused interventions linked to dynamic assessment. As researchers and clinical speech and language therapists working with deaf children in a variety of health and education settings, their shared expertise has helped take research-based tools and adapt them for use in functional settings with children. Additionally, this work has helped identify ways to support others to use dynamic assessment methods and shown steps for future work.

When assessing deaf children and their use of signed language, consideration of the individual's strengths and needs is paramount. Two cases described in a research study (Mann, Peña, & Morgan, 2014) highlight how two children with below average scores on a standardised American Sign Language test can have very different language learning difficulties, and therefore, different support needs. When teachers and speech and language therapists meet such children in their everyday practice, their understanding of the child's language needs and context will underpin intervention and progress.

1.1 Deaf children

As with all children, deaf children have their individual needs and strengths. Some deaf children develop language following a typical development path, however, a sub-population of deaf children may have additional needs linked to the cause of their deafness (Inscoe & Bones, 2016). For many children, the diagnosis of deafness is unexpected by parents. Approximately 95% of deaf children are born into families with no history or experience of deafness (Mitchel and Karchmer, 2004). Currently, for children diagnosed as deaf at an early age, the initial contact parents have with services is through audiology staff. This sets parents on a path dominated by the 'medical model' of deafness, where hearing loss is viewed as a disability that needs to be 'fixed' and all focus is on minimising or eliminating an inability to hear (Young et al., 2006; Young et al., 2005). Consequently, hearing-aid and cochlear implant centres may not place their primary focus on the language development of children but, instead, on their audiological management (Humphries et al., 2012). Although parents have the right to decide on the pathway of intervention they follow for their children, the lack of information or access to services often makes a truly informed choice difficult for them. For the parents of children with complex needs such as autism, learning disability or sensory processing disorder, it can be challenging to understand how parental choices during the early years impact on the child's language development within what is considered the 'critical language learning period'.

1.2 Learning language as a deaf child

Deaf children, like all children learning language, have a period when their brain is more receptive to language learning opportunities (Twomey, et al 2020; Veríssimo, et al 2018). Depending on parental knowledge of a signed language, deaf children may have no exposure to a signed language in their early years, and a child's access to spoken language will be dependent on their hearing and listening skills (Marriage, Brown, & Austin, 2017), where hearing levels indicate the sounds a child can perceive and listening skills describe the sounds a child can understand and attribute meaning to. For the 5% of deaf children who are born into a family with experience of deafness, a signed language may provide full access to a language model from birth and through the aforementioned critical language learning period. For parents who choose to learn a signed language after their child has been diagnosed as deaf, it means that they will be learning a language alongside their child and being their role model although such role models are not equivalent to a (deaf) parent who is a fluent signer (Lu, Jones, & Morgan, 2016). Some children who have exposure to a signed language in their early years may only see language used by their immediate family. Other deaf children who use a signed language may have access to peers, family members and a community who sign if they are born as part of the Deaf community or if their parents make choices that allow this kind of access.

1.3 Deaf community

In the UK Deaf community, British Sign Language (BSL) is the dominant language for everyday communication. The British Deaf Association estimates 87,000 people prefer to use BSL and 151,000 people use BSL at home (*https://bda.org.uk/help-resources/* accessed 24.12.2020). This second figure includes people such as hearing family members of Deaf people. Members of the Deaf community have a strong sense of identity linked to their Deafness, the history of Deaf people, and their status as a minority community (Ladd & Lane, 2013).

1.4 Deaf community language culture

In the UK, there are mainstream schools which the majority of children attend, hearing and deaf. There are local day special schools for various learning difficulties and, relatively few, residential special schools (*https://www.batod.org.uk/information/special-schools-deaf-children-uk/accessed* 29.12.20) which serve regions of the country. Residential special schools for deaf children have seen a declining trend over recent years due to the government's philosophy of inclusion in mainstream schools for all (*https://www.batod.org.uk/information/the-education-of-deaf-pupils-in-mainstream-schools/accesses* 29.12.20), cochlear implantation as a 'fix' for deafness, and the costs to local councils of educating a child in a residential specialist facility. However, residential special schools for deaf children have been acknowledged for many years as strong deaf communities in themselves both linguistically and culturally (Ladd, 2018). They offer deaf children the opportunity to learn and communicate in the language of the wider deaf community in the UK: BSL, and support children to learn about the rich heritage and day to day lives (and struggles) of deaf people. Identity and community are very important to mental health. Everyone needs a language, a means of communication and accessible conversation partners.

Mainstream schools or local special schools, may offer 'sign' but this is often Sign Supported English or Makaton: signs that are used alongside spoken English and not the same as BSL which has its own vocabulary and grammar, and is fully visual-spatial. BSL does not rely on listening at all.

2. A practitioner's dilemma

When working with a deaf child, practitioners may find that language assessment is problematic. This could be for a number of reasons, one being the great majority of children growing up in hearing families and showing considerable variability in their language experience, as explained above. Another reason is that deaf children are traditionally assessed on standardised assessments for spoken language (there is a global shortage of assessments developed specifically for signing deaf children). Perhaps unsurprisingly, they tend to perform low on these assessments. As a consequence, many deaf children may be misdiagnosed or over diagnosed as having a language impairment whereas, in fact, their low performance may be due to a language difference – after all, the test that is used was developed for and normed on a different population: hearing children. This leaves the practitioner with questions on how to differentiate between a deaf child's low test performance due to:

- language learning problem?
- language difference?
- language delay?
- language deprivation, or
- other cognitive issues?

For test administrators it is important to make this distinction in order to avoid a mismatch between the support that is recommended for the deaf child and the child's actual needs.

3. Dynamic assessment

Dynamic assessment (DA) combines teaching and assessment processes within a single assessment procedure to measure learning potential and evaluates the enhanced performance that results from learning through mediation. DA is an umbrella term for a range of different approaches used in Psychology and Education that blend teaching and assessment into a single assessment approach with the aim to measure a child's learning potential or modifiability (the ability to carry over newly learned skills and the amount of support required) and to evaluate the enhanced performance that results from mediated learning. The most common approaches include test-teach-retest and graduated prompting. DA draws largely on the ideas of Russian psychologist Lev Vygotsky who proposed that a child learns best through social interaction and can develop higher mental functioning through collaboration and interaction with a more experienced peer or adult. His ideas have been applied by others (Feuerstein, 1979; Lidz, 1987, 1991) when describing the mediation interaction (Mediated Learning Experience, MLE) that occurs during teaching phases of dynamic interaction (Peña, Iglesias & Lidz, 2001). Many approaches in DA are motivated by the aim to develop instruments that can provide practitioners with a short and structured way to determine how a child may respond to a specific type of support in the future. DA itself is not intervention.

DA of deaf children is still a very young field with two main strands of research. The first strand of research was carried out during the 90s-early 2000s with focus on cognitive skills. It was motivated by deaf children's low performance on standardised IQ tests (e.g., Lidz,

2004; Tzuriel & Caspri, 1992). The other, more recent, strand has explored the use of DA within a language learning context. This includes research on deaf children's narrative skills in English (Asad, Hand, Fairgray, & Purdy, 2013), on signing deaf children's vocabulary knowledge (Mann, Peña, & Morgan, 2014, 2015), and on enabling practitioners to understand and use DA with children who have atypical language development (Hoskin, 2017; Marshall et al., 2020).

Moving DA from research to practice requires work not only with the practitioner but also the child's family. Both may need support to observe and understand how this type of assessment will benefit the child and themselves.

Intervention focuses on identifying the skills a child has in language, attention, engaging with others and understanding tasks. It then supports those interacting with the child to select strategies that suit a specific child and support their learning and self-management. For more detailed information on how DA can inform intervention see Mann et al (in press).

4. Case studies

The following two brief case studies shall demonstrate the practical application of DA.

4.1 Case study 1

A deaf boy, who came to the UK aged fourteen, presented with complex language needs in school. Using dynamic assessment techniques and gathering information from different communication partners, his language skills and needs were better understood. He had learnt a little of his home country's spoken language and a little of his country's signed language. On arriving in the UK, he learnt a little spoken English. On arrival in school, he was just beginning to learn BSL. He was confused about his situation and, along with some adults, didn't realise he was having experience of four languages between home and school. During lessons, he needed regular support to understand that his situation was complex, to give himself credit for his achievements and to know which of the four languages he was working in at any given time. When his language context and current skills were understood by those working with him, it was possible to use strategies that identified which language he was using, whether he had other language or cognitive skills that linked to the language required in the current task, and whether using these skills was helpful.

4.2 Case study 2

A deaf girl aged twelve was identified as having difficulties with language. During language assessment using a DA approach, it was noted that the child had very low self-esteem, needing support to manage their internal emotional state. Strategies that supported her to do this included being allowed to be 'the teacher'. By switching roles, her control over the language situation was increased. When presented with a mediator who said 'Pretend you're the teacher and I'm you, here's the work, what do I need to do', the child was able to demonstrate language and cognitive strategies to describe the task and the action that was needed. She was able to use her languages skills, problem solve, engage with the mediator and complete tasks because of this change in strategy.

4.3 Summary of why dynamic assessment was useful in these cases

Dynamic assessment was useful in working with these children because:

- It combines teaching and assessment, allowing the teacher to assess the child's skills and their own strategy-use within one activity in an explicit way
- It is complementary to standardised assessments, not replacing them but providing additional information about a child's learning capacity and style
- It can be very useful for children who represent a puzzle to practitioners, providing a breakdown of adult and child strategies to consider within a dynamic assessment and mediated learning environment framework
- Where a child appears to be non-responsive to every-day-teaching strategies, dynamic assessment can help a practitioner consider their own strategies, and the impact these have on the child's ability to change their participation and language use

5. Next steps

To make DA techniques available to a wider group of practitioners, a close collaboration between researchers and practitioners is essential. We have identified four key areas for focus.

5.1 Information and tools

When evidence of the effectiveness of DA with children, is reported in research papers, the concepts and materials reported on, often need to be adapted by researchers and practitioners for deaf children and their specific contexts. These tools can then be used in training courses and peer support groups, making sure research in DA is accessible to practitioners who work in signed languages.

5.2 Understanding dynamic assessment for measuring progress

The historic reliance on standardised assessment to measure children's progress is changing. Some schools are now taking an ipsative referenced approach. Providing advice for practitioners on how to use feedback from DA in annual reviews and reports will be needed. Explanations of why this type of measurement is valid and reliable, how it can be trusted, and advice on sharing a culture within a school or other establishment for the use of dynamic assessment is needed.

5.3 Including the child in the assessment process

As described in the case studies DA and MLE techniques can be used to include a child in discussions about their learning strengths, needs and challenges. Accessible information and tools which support the inclusion of the child in this process are available but need to be shared more widely.

5.4 Training

For practitioners who have BSL as their dominant language, specific training opportunities are needed. An Erasmus + funded project, DOTDeaf, is currently working to provide accessible information for co-learning between English and BSL-using practitioners. The training course being developed includes some adapted DA and MLE tools (See - https://blogs.city.ac.uk/dotdeaf/).

DA techniques help practitioners focus on connecting with a child in a way that enables the adult and child to work together, using the child's strengths. Information and training about the use of these techniques needs to be available, including for practitioners who use BSL. DA can help practitioners share information to help family, staff and others understand and use the potential of each individual child for language learning. Practitioners need support and training in order to do this.

6. Conclusion

Standardized language tests are useful but can underestimate deaf children's true abilities. DA offers an alternative, yet complementary, approach to standardized testing. DA makes the assessment process more interactive and fun, enabling the assessor and child to try out different processes, and also freeing the assessor to observe carefully. DA can empower the child to show their strengths, preferences and motivators so that the assessor can discover the circumstances that enable the child to learn optimally. The assessor and child are collaborating in uncovering and setting up the best learning situations, and together they can impart this new knowledge about the child to others in the child's world including teachers and parents/carers. DA can be used in an ongoing way to monitor progress in learning. It is an attitude to assessment and learning which looks very closely at what works for the individual child, aims to provide what is needed, and is deeply respectful of the child's contribution to the learning partnership.

References

Asad, a. N., Hand, L., Fairgray, L., & Purdy, S. C. (2013). The use of dynamic assessment to evaluate narrative language learning in children with hearing loss: Three case studies. *Child Language Teaching and Therapy*, 29(3), 319–342. https://doi.org/10.1177/0265659012467994

Bruce, B., Thernlund, G., & Nettelbladt, U. (2006). ADHD and language impairment: A study of the parent questionnaire FTF (Five to Fifteen). *European Child & Adolescent Psychiatry*, 15(1), 52–60. https://doi.org/10.1007/s00787-006-0508-9

Feuerstein, R. (1979). T*he dynamic assessment of retarded performers.* Baltimore, MD: University Park Press.

Hoskin, J. H. (2017). *Language Therapy in British Sign Language: A study exploring the use of therapeutic strategies and resources by Deaf adults working with young people who have language learning difficulties in British Sign Language (BSL).* Unpublished doctoral thesis. UCL (University College London).

Humphries, T., Kushalnagar, P., Mathur, G., Napoli, D. J., Padden, C., Rathmann, C., & Smith, S. R. (2012). Language acquisition for deaf children: Reducing the harms of zero tolerance to the use of alternative approaches. *Harm Reduction Journal*, 9(1), 16. https://doi.org/10.1186/1477-7517-9-16

Inscoe, J., & Bones, C. (2016). Additional difficulties associated with aetiologies of deafness: outcomes from a parent questionnaire of 540 children using cochlear implants. *Cochlear Implants International*, 171(1), 21–30. https://doi.org/10.1179/1754762815Y.0000000017

Ladd, P. (2018). *Understanding Deaf Culture.* https://doi.org/10.21832/9781853595479

Ladd, P., & Lane, H. (2013). Deaf Ethnicity, Deafhood, and Their Relationship. *Sign Language Studies*, 13(4), 565–579. https://doi.org/10.1353/sls.2013.0012

Lidz, C. S. (2004). Successful application of a dynamic assessment procedure with deaf students between the ages of four and eight years. *International Journal of Theoretical and Applied Finance*, 7(2), 59-73.

Lidz, C. S. (1991). *Practitioner's guide to dynamic assessment.* Guilford Press.

Lidz, C. S. E. (1987). *Dynamic assessment: An interactional approach to evaluating learning potential.* Guilford Press.

Lu, J., Jones, A., & Morgan, G. (2016). The impact of input quality on early sign development in native and non-native language learners. *Journal of Child Language*, 43(03), 537–552. https://doi.org/10.1017/S0305000915000835

Mann, W. Hoskin, J. & Dumbrill, H. (in press). Using dynamic assessment to assess the langauge and communication skills of signing deaf children. in T. Haug, W. Mann, & U. Knoch (Eds.), *Handbook of language assessment across modalities.* OUP.

Mann, W., Peña, E. D., & Morgan, G. (2015). Child modifiability as a predictor of language abilities in deaf children who use American Sign Language. *American Journal of Speech-Language Pathology*, 24(3), 374-385.

Mann, W., Peña, E. D., & Morgan, G. (2014). Exploring the use of dynamic language assessment with deaf children, who use American Sign Language: Two case studies. *Journal of Communication Disorders*. https://doi.org/10.1016/j.jcomdis.2014.05.002

Marriage, J., Brown, T. H., & Austin, N. (2017). Hearing impairment in children. *Paediatrics and Child Health*, 27(10). https://doi.org/10.1016/j.paed.2017.06.003

Marshall, C., Rowley, K., Atkinson, J., Denmark, T., Hoskins, J., & Sieratzki, J. (2020). Chapter 5. *Atypical sign language development* (pp. 73–92). In G. Morgan (Ed.).

Understanding Deafness, Language and Cognitive Development (Vol 25). Amsterdam: John Benjamins Publishing. https://doi.org/10.1075/tilar.25

Mitchell, R. E., & Karchmer, M. (2004). Chasing the mythical ten percent: Parental hearing status of deaf and hard of hearing students in the United States. *Sign language studies*, 4(2), 138-163.

Peña, E., Iglesias, A., & Lidz, C. S. (2001). Reducing test bias through dynamic assessment of children's word-learning ability. *American Journal of Speech-Language Pathology*, 10(2), 138-154.

Twomey, T., Price, C. J., Waters, D., & MacSweeney, M. (2020). The impact of early language exposure on the neural system supporting language in deaf and hearing adults. *NeuroImage*, 209(June 2019). https://doi.org/10.1016/j.neuroimage.2019.116411

Tzuriel, D., & Caspi, N. (1992). Cognitive modifiability and cognitive performance of deaf and hearing preschool children. *The Journal of Special Education*, 26(3), 235-252.

Veríssimo, J., Heyer, V., Jacob, G., & Clahsen, H. (2018). Selective Effects of Age of Acquisition on Morphological Priming: Evidence for a Sensitive Period. *Language Acquisition*, 25(3), 315–326. https://doi.org/10.1080/10489223.2017.1346104

Young, A., Carr, G., Hunt, R., McCracken, W., Skipp, A., & Tattersall, H. (2006). Informed choice and deaf children: underpinning concepts and enduring challenges. *Journal of Deaf Studies and Deaf Education*, 11(3), 322–336. https://doi.org/10.1093/deafed/enj041

Young, A., Jones, D., Starmer, C., & Sutherland, H. (2005). Issues and dilemmas in the production of standard information for parents of young deaf children - Parents' views. *Deafness and Education International*. https://doi.org/10.1002/dei.27

Index

A

Aboriginal, 22
ACC, 15
Adams, 8
Aegean, 44
Aissen, 14, 16
Alan, 4, 7, 8, 37, 41
Albanian, 48
Alevi, 48
Ama, 51
American, 66, 73, 74
Anderson, 7, 21, 23, 28
Apulian, 15
Ara, 45, 50, 51, 52
Arabic, 22, 46
Armenian, 46
Armidale, 43
Aryan, 23
Asad, 70, 72
Asian, 22
Atlas, 11, 16
Attila, 58, 63
Austin, 67, 73
Australia, 4, 7, 8, 9, 18, 19, 20, 21, 22, 23, 27, 28, 29, 30, 31, 41, 43, 53
Australian, 20, 22, 28, 29, 32, 41, 42

B

Bakhtygul, 4, 8
Barac, 22, 24, 27, 28
Barger, 55, 63
Basilaia, 32, 41
Bektashi, 48
Berruto, 12, 16
Bialystok, 20, 21, 22, 24, 27, 28, 29
Bohm, 7
Borjigid, 7
Bossong, 15, 16
Brighton, 65
British, 68, 73
Brooke, 40, 41
Brown, 67, 73
Bruin, 21, 28, 29
Bruni, 7
Bucharest, 7, 8
Bulgaria, 44, 50
Bulgarian, 44, 46
Buonocore, 7, 9

C

Calabrian, 15
Cambodian, 23
Cantonese, 22
Caspri, 69
Chambers, 12, 16
Charles, 8
Chi-Square, 59, 61
Chomsky, 45, 52
Chomskyan, 45
Christo, 41
Clyne, 22, 28
College, 8, 65, 73
Coluzzi, 13, 16
Comrie, 15
Cope, 33, 42
COVID-19, 7, 8, 9, 10, 11, 13, 15, 16, 30, 31, 34, 41, 42

Creswell, 35, 41
Croft, 15
Crowe, 8
Crystal, 10, 16
Czech, 8

D

Danica, 7
Davidson, 21, 28
Deaf, 8, 65, 66, 67, 68, 73, 74
Devrim, 8, 43
Dim, 33, 40, 41
Donato, 8
Dorcas, 8
Dumbrill, 8, 65, 73

E

Eastern, 44, 52
Edirne, 45, 49, 50
ELICOS, 8, 30, 31, 34, 35, 42
Empire, 45
Empungan, 33, 42
England, 8, 43
English, 20, 21, 22, 23, 28, 31, 34, 36, 39, 40, 41, 42, 48, 49, 53, 63, 68, 70, 72
Erasmus, 72
Europe, 13, 44, 52
European, 11, 22, 42, 44, 73

F

Faculty, 53
Farsi, 46
Filipino, 23
Florentine, 12
Fotos, 41

G

Gagliastro, 8
Galambos, 20, 21, 29

Gelibolu, 45
German, 22
Germanic, 21, 23, 25, 26
Giacomo, 7
Glides, 48
Goldin, 20, 21, 29
Goldsmith, 55, 63
Grandey, 55, 63
Greece, 44, 50
Greek, 8, 22, 23, 44, 46, 48
Gross, 55, 63
Gullifer, 25, 27, 29

H

Hakuta, 21, 29
Halliday, 45, 52
Hamilton, 8, 65
Hebrew, 46
Hilary, 8, 65
Ho, 15
Holtus, 12, 16
Hoop, 15
Horwitz, 33, 40, 42, 63
Hoskin, 8, 65, 70, 73

I

Iceland, 8
India, 41
Indo, 23
Iranic, 23
Iryna, 7, 18
ISTAT, 10, 12
Italian, 10, 11, 12, 13, 14, 15, 16, 22
Italy, 11, 12, 13, 14, 16

J

January, 57, 58
Jessner, 20, 29
Joanna, 8, 65
Johnson, 55, 63

Jones, 67, 73, 74

K

Karchmer, 67, 74
Kathryn, 8
Kazakh, 8
Kemal, 45, 52
Khodos, 7, 18, 21, 27, 28, 29
Kinstler, 57, 63
Koomson, 31, 42
Koondhar, 32, 42
Krumsvik, 33, 34, 42
Kuala, 58
Kvavadze, 32, 41

L

Ladd, 68, 73
Ladino, 46, 48
Laine, 21, 28, 29
Lambert, 20, 29
Latin, 8, 49
Lehtonen, 21, 28, 29
Lepschy, 11
Lev, 69
Libert, 4, 7, 8, 41
Lidz, 69, 73, 74
Likert, 23
Litvin, 55, 63
London, 8, 16, 52, 65, 73
Lu, 67, 73
Lumpur, 58

M

Macedonian, 48
Magdolna, 4, 7, 8
Maiden, 12, 16
Makaton, 68
Makhanbetova, 4, 8
Malay, 23
Malayalam, 23
Malaysia, 8, 53, 54, 55, 58, 63

Malchukov, 15
Mandarin, 22
Mann, 8, 65, 66, 70, 73
March, 14, 57, 58
Margaret, 8
Maria, 4, 7, 8
Mario, 15
Marmara, 44
Marshall, 70, 73
Martin, 28, 54, 56, 57, 60, 63
Master, 22
Mengshan, 8, 30
Merriam, 49, 52
Microsoft, 35
Middle Eas, 22
Mitchel, 67
MLE, 69, 72
Moorhouse, 31, 42
Moreno, 22, 24, 29
Morgan, 66, 67, 70, 73
Moskovsky, 21, 27, 29, 41
Moso, 7
Moss, 11

N

Naabuyun, 47
Neapolitan, 10, 13, 15
Newcastle, 4, 7, 8, 9, 18, 21, 30, 31, 34, 41, 53
Nordin, 33, 42
Northern, 44
November, 1, 3, 4, 7, 8
NVivo, 58

O

OneDrive, 31, 35, 36, 37, 38, 40
Oslo, 4, 7, 8
Ottoman, 45
Owusu, 31, 42

P

Pahang, 58
Pan, 55, 63
Patton, 58, 63
Picard, 54, 63
Pomak, 48
Prague, 8
Primus, 15
Psychology, 69
Purdy, 70, 72

R

Radtke, 12, 16
Republic, 8, 45
Robinson, 54, 63
Rodney, 8
Roehampton, 8, 65
Romanca, 48
Romanian, 15, 16
Romanic, 23
Rubino, 22, 29
Rushton, 8
Russian, 69
Ryan, 20, 24, 28

S

Saleh, 33
Salehi, 33, 42
Sarah, 8, 30
SARS, 31, 41
Sato, 32, 42
Savignon, 32, 40, 42
Selangor, 58
Shah, 33, 42
Shona, 23
Sicilian, 13, 15
Siming, 32, 42
Sinic, 23
Slavic, 7, 23
South, 7, 28, 52
Southeast, 44
Southern, 10, 11, 14, 15
Spanish, 15, 16
Spencer, 55, 56, 63, 64
Steve, 8

T

Tatar, 7, 8
Tekirdaaya, 49
Tekirdağ, 45, 49
Thailand, 54
Thaw, 55, 63
Tibeto, 23
Tomlinson, 33, 42
Trakya, 8, 43, 44, 45, 47, 48, 49, 50, 51
Trakyaca, 44, 45, 46, 47, 48, 49, 50, 51
Trudgill, 12, 16
Turkey, 44, 45
Turkic, 45
Turkish, 44, 45, 46, 49, 51, 52
Turkology, 45
Twomey, 67, 74
Tzuriel, 69, 74

U

UK, 8, 22, 63, 65, 68, 70
Umrani, 32, 42
UNESCO, 11, 12, 16, 31
University, 4, 7, 8, 16, 18, 28, 29, 30, 31, 32, 34, 41, 43, 53, 65, 73

V

Vanetti, 55, 63
Vaughn, 25, 29
Venetian, 13
Verette, 15
Veríssimo, 67, 74
Vietnamese, 22, 23, 42
Visti, 15

Vitti, 15
Vygotsky, 20, 29, 69

W

Wang, 55, 64
Wen, 8, 53
Western, 44
White, 54, 56, 57, 60, 63
Wilson, 52
Wolfgang, 8, 65
Word-of-Mo, 54

X

Xiamen, 7

Xu, 8, 30

Y

Yılmaz, 43
Youtube, 35
Yu, 33, 42
Yunus, 33, 42

Z

Zealand, 31
Zoom, 7, 31, 34, 35, 36, 37, 38, 39, 40
Zulkiffli, 55, 63
Zuvalinyenga, 8

www.ingramcontent.com/pod-product-compliance
Lightning Source LLC
Chambersburg PA
CBHW060435220526
45465CB00008B/3149